ANTHROPOLOGICAL PAPERS

MUSEUM OF ANTHROPOLOGY, UNIVERSITY OF MICHIGAN
NO. 23

ABORIGINAL RELATIONSHIPS BETWI CULTURE AND PLANT LIFE IN THE UPPER GREAT LAKES REGION

by
RICHARD ASA YARNELL

ANN ARBOR
THE UNIVERSITY OF MICHIGAN, 1964

ISBN (print): 978-1-949098-22-8
ISBN (ebook): 978-1-951519-45-2

Browse all of our books at
sites.lsa.umich.edu/archaeology-books.

Order our books from the University of Michigan
Press at www.press.umich.edu.

For permissions, questions, or manuscript queries,
contact Museum publications by email at umma-
pubs@umich.edu or visit the Museum website at
lsa.umich.edu/ummaa.

PREFACE

This report deals with interrelationships between aboriginal culture and plant life in the Upper Great Lakes region with greatest concentration on (1) utilization of native plant products as determined from ethnographic records and the results of archaeological research; and (2) aboriginal agriculture with respect to its historical aspects and the distribution of length of frostless season. In addition, it has been possible to gather information and to make certain interpretations about the effects of aboriginal activities on the natural flora and vegetation of the Upper Great Lakes region, primarily those involving subsistence activities.

The ethnographical data on the uses of plants were collected from unpublished works and manuscripts as were most of the archaeological records of cultrivated plants. However, with the generous assistance of Volney H. Jones and Edward G. Voss, I located and identified, in the laboratory, most of the archaeological remains of uncultivated plant food and charcoal of the Upper Great Lakes region. Data on the natural distribution and habits of the plants involved were collected primarily from the files of Dr. Voss and from published works supplemented by my observations in the field, interpreted with the generous assistance of William S. Benninghoff. Various field trips were conducted to archaeological sites, some of which were undergoing excavation at the time by field crews of the University of Michigan Museum of Anthropology during the summers of 1960 and 1961. These included visits to Isle Royale, to Bois Blanc Island, on three occasions; to sites in the Saginaw area, on many occasions; and to various other sites in Lower Michigan.

All of the collections, files, reports, and library of Volney H. Jones and the laboratory of ethnobotany were made available and furnished a large amount of the information needed for this study. The unpublished ethnobotanical reports and archaeological plant remains were especially valuable. The Museum of Anthropology library, collections, and equipment in general were made available through the courtesy of James B. Griffin, the director. These materials were invaluable in the prosecution of the investigation. In addition, many of the field trips and much of the time spent in laboratory research were made possible by financial support from the Museum of Anthropology.

I wish to acknowledge with gratitude the financial support received for this work from the National Science Foundation grant to the University of Michigan Museum of Anthropology for research in "A Correlation of Prehistoric Cultural Complexes and Post-Pleistocene Ecologies in the Upper Great Lakes," under the direction of James B. Griffin. Appreciation is also due the Homeopathic Hospital Guild Scholarship administered by the University of Michigan Museum of Anthropology for financial support of research on the utilization of native plants by the Indians of Michigan, and to the Horace H. Rackham School of Graduate Studies for support in the form of a summer fellowship for writing much of this study. In addition, I am greatly indebted to Professors Volney H. Jones, William S. Benninghoff, Edward G. Voss, Arthur J. Jelinek, James B. Griffin, Emerson F. Greenman, and Hugh C. Cutler for advice and encouragement; and to Meredith J. Black, Charles E. Cleland, Richard E. Flanders, Alan L. McPherron, Mark R. Papworth, G. Richard Peske, William B. Roosa, Albert T. Steegman, David W. Taggart, Richard G. Wiegert, and Henry T. Wright for their assistance. Special thanks is due George W. Stuber for the contribution of his patience and skill in map-making and in photography.

Richard Asa Yarnell

CONTENTS

I

INTRODUCTION

The purpose of this investigation is three-fold: (1) to collect ethnographic and archaeological data bearing upon the relationships between plants and the aboriginal cultures of the Upper Great Lakes region; (2) to make certain interpretations of the data available at the present time; and (3) to give preliminary indications of the kinds of data that can be made available and the kinds of interpretation that can be made in studies of aboriginal plant—man relationships in this and in adjacent regions, especially as they have bearing upon subsistence.

The data and interpretations break rather easily into two categories which overlap only slightly in the region of study, though the overlap apparently increases toward the south. One category involves plants that were introduced by man from the south and survive in the region through the agency of man, that is, most of the plants intentionally cultivated. The other category involves plants that are presumably native and that survive and reproduce without the interference of human activities but whose products are collected and utilized by man.

Chapters II and III deal primarily with plants in the second category—the "native" plants. Chapters V and VI deal with the plants that survive and reproduce as a result of human activities, that is, the cultigens. Chapter IV deals with the kinds of relationships that seem to exist in the areas of overlap.

Physical Character of the Area

The Upper Great Lakes region is here defined as the area that drains into Lakes Huron, Michigan, and Superior. It includes almost all of Michigan, the northern edge of Indiana, the northeast corner of Illinois, the northeast part of Wisconsin, the northeast edge of Minnesota, and that part of Ontario bordering on Lake Superior and Lake Huron. The entire drainage basin covers an area of 220,480 square miles with 77,230 miles of shore line (see Table I for statistics for each lake).

Many smaller lakes are also present in the area. Michigan alone has at least 4187 lakes covering 500 square miles

TABLE I

SOME STATISTICS OF THE UPPER GREAT LAKES (MARKS, 1961: 26)

	Great Lakes	Upper Great Lakes	Lake Huron	Lake Michigan	Lake Superior
Entire drainage basin (sq. mi.)	295,200	220,480	72,620	67,860	80,000
Water surface area (sq. mi.)	95,170	77,230	23,010	22,400	31,820
Shore line (mi.)	10,977	7,822	3,185	1,661	2,976
Length (mi.)	.....	.....	206	307	350
Breadth (mi.)	.....	.....	101	118	160
Maximum depth (ft.)	.....	.....	750	923	1,333
Average depth (ft.)	.....	.....	195	276	487
Mean surface elevation (ft.)	.....	.....	580.54	580.56	620.20
Average annual precipitation (in.)	31	30	31	31	29

(Hinsdale, 1932: 15), but formerly the number was much larger because it included many lakes which have since become marshes, swamps, and bogs. These lakes were formed as a result of the Wisconsin ice advance which reached its southernmost limit approximately 20,000 years ago. Since that time the present character of the Great Lakes and their drainage systems have evolved. This process was essentially completed by about 9000 years ago at which time the ice had retreated north of the region, though various fluctuations of lake level and area and of drainage pattern have taken place since then, especially during the Chippewa-Stanley stage, when lake levels were much lower than they are now.

Since the Wisconsin ice retreated, the climate has become warmer, with the warmest and driest conditions prevailing during the hypsithermal period which seems to have reached its climax about 3000 to 4000 years ago. Since then there has been a cooling trend that has continued to the present, but with various interruptions (Griffin, 1960*a*).

The present climate varies according to latitude and proximity to the lakes. Trewartha (Espenshade, 1962: 6) assigns most of the region to a climatic zone that is humid continental with cool summers, and is constantly moist with precipitation throughout the year. The northern fringe is subarctic, but warm summers are characteristic of the southern fringe. Despite its inland position, however, the region has a climate that is not altogether continental in nature because of the moderating influences of the Great Lakes (see Chapter VI for further discussion of climate).

Soils are derived mostly from glacially deposited and lake deposited parent material. Podsol soils are predominant in the north, but bog soils are common in low areas, and lithosoils and shallow soils exist in some of the higher areas around Lake Superior. Gray-brown podsolic soils predominate in the south except in some of the lower areas where ground-water podsol and half-bog soils are present, especially in the Saginaw Basin (H. G. Knight *in* Williams, 1959: 47).

The Upper Great Lakes region ranges in elevation from 580 feet along the shores of Lake Huron and Lake Michigan to over 2000 feet in northeast Minnesota, but very little of the land is above 1600 feet, and most of that is south and west of Lake Superior. Otherwise most of the region is relatively level or rolling country averaging below 1000 feet in elevation with surface features that are typical of lands subjected to continental glaciation. Thus there are no barriers except the lakes themselves, but even these features apparently served more as highways than as barriers in good weather, in combination with the various streams suitable for canoe travel.

Michigan alone has thirty-four primary river systems and 3,177 miles of coastline. "Michigan's numerous streams are, in general, short. On the Lower Peninsula, they flow gently, smoothly over well-worn beds; only a few are turbulent and scurrying. The rivers of the Upper Peninsula, however, are turbulent and scurrying" (Michigan Writers' Program, 1941: 18).

The mineral resources of interest to the Indian included various kinds of rock for use in the manufacture of tools and weapons, especially flint, chert, and argillite. These materials are available primarily as nodules scattered in the glacial debris. Native chert, however, is available around Saginaw Bay. Of more consequence, perhaps, are the deposits of native copper in western Upper Michigan and Isle Royale which were mined for thousands of years by the Indians who traded this metal widely because of its scarcity and utilitarian value. Abundant clay deposits were utilized for pottery manufacture during the last 2000 years of aboriginal occupation.

Vegetation

The natural vegetation of the Upper Great Lakes region is composed mostly of forests of four major types: oak-hickory (central hardwoods), birch-beech-maple-hemlock (northeastern hardwoods), jack, red, and white pines (northeastern pine forest), and spruce-fir (northern coniferous forest). The oak-hickory forest occupies most of southern Lower Michigan, except near Lakes Michigan and Huron, and extends northward in the Saginaw region to Saginaw Bay and eastward into the southern tip of Ontario. The pine forests occupy large tracts in north-central Lower Michigan, Upper Michigan, northern Wisconsin, northern Minnesota, and southern Ontario. The spruce-fir forest occurs mostly through the same range as the pine forest—in bogs and swamps and in the highlands of Upper Michigan, but toward the north it becomes the prevalent forest type. Otherwise the northeastern hardwood forest prevails, occupying nearly 25,000,000 acres in the Lake states (Shantz and Zon, 1924).

The exceptions to this are minor. These are marshes, primarily in the Saginaw area; and bogs, swamps, and small isolated prairie areas in the southwest. Other vegetation types of any importance, except for southern swamplands, are the result of recent human activities. It may be that the nature of some of the vegetation is partly the result of former Indian activities, especially in the case of the southwestern Michigan grassy openings in the forest; but this would be difficult to demonstrate.

G. E. Nichols (1935) refers to the region of the northeastern hardwoods as "the hemlock - white pine - northern hardwood region or simply "the eastern hemlock region." He depicts it (map, p. 405) as extending from northern Minnesota and northern Wisconsin through Upper Michigan and most of Lower Michigan to the lower St. Lawrence Valley, Nova Scotia, and New England with a narrow southern extension down the Appalachian Mountains into eastern Tennessee.

Nichols (1935: 404-7) characterizes the eastern hemlock region in relation to the forest adjacent to the north and to the south:

> . . . it is flanked along its northern borders by the vast region of northern coniferous forest. . . . Toward the south it impinges on the region of deciduous forest which occupies much of the eastern United States. . . . The northern conifer forest, monotonously uniform over vast stretches of country, is impressive in its simplicity. It comprises scarcely half-a-dozen dominant tree species, chief among these being the balsam fir, *Abies balsamea*, white spruce, *Picea glauca*, black spruce, *Picea mariana*, and paper birch, *Betula papyrifera*. . . . Small trees and shrubs are comparatively few in number. . . . The deciduous forest climax, on the other hand, is highly complex in its structure, varying greatly from place to place. It comprises a great wealth of tree species, all of which are deciduous—the so-called "hardwoods." Prominent among these are numerous oaks, *Quercus* spp., and hickories, *Hicoria* spp., chestnut, *Castanea dentata*, tulip poplar, *Liriodendron tulipifera*, beech, *Fagus grandifolia*, sugar maple, *Acer saccharum*, white ash, *Fraxinus americana*, black cherry, *Prunus serotina*, basswood, *Tilia* spp. . . . and black walnut, *Juglans nigra*. . . . The forest tends to be several storied, with one or more layer societies of shrubs and small trees.
>
> Forest conditions in the eastern hemlock region, as might be expected, are transitional in many ways between those of the coniferous forest region to the north and those of the deciduous forest region to the south. Large areas of country are occupied by evergreen coniferous forest and large areas by deciduous hardwood forest, while other large areas are forested with a mixture of conifers and hardwoods. But the forest vegetation also exhibits certain features which are both unique and distinctive. As a result there exists not a little confusion or difference of opinion as to how this region should be treated in relation to that of the two adjoining ones.

L. R. Dice (1938) agrees that the northeastern hardwood forest region is distinctive, but he adds a faunal dimension to it and refers to it as the "Canadian Biotic Province" which he defines as "that part of northeastern North America in which hardwoods form the climax and conifers of several kinds form several types of subclimaxes" (pp. 503-5).

Dice (1938:508) characterizes the vegetation of the Canadian Biotic Province as follows:

> The climax vegetation of the Canadian province is a hardwood forest, in which the sugar maple (*Acer saccharum*) and yellow birch *(Betula lutea)* are the most characteristic trees. The eastern white pine *(Pinus strobus)* and the eastern hemlock *(Tsuga canadensis)* occur frequently. The beech is also characteristic, except that it does not occur in the extreme western part of the province. Numerous other trees and many shrubs are characteristic. . . .
>
> The pines . . ., often growing in a pure stand of one species, are in many places an important subclimax stage in the Canadian province. The pine forests nearly always occupy sandy or gravelly soil. On some very poor sandy soils the pine forest may persist indefinitely as an edaphic climax, for in such situations insufficient humus may accumulate ever to make the site suitable for a hardwood forest. In poorly drained situations in the Canadian province, many areas are covered by bogs and swamps in which there occur varying mixtures of balsam fir *(Abies balsamea)*, black-spruce *(Picea mariana)*, white spruce *(Picea canadensis)*, northern white cedar *(Thuja occidentalis)*, and tamarack *(Larix laricina)*, with occasional hardwoods such as black ash and red maple.

J. E. Potzger and others refer to the forest that extends from the Great Lakes to New England as the "Lake Forest Formation." In the western portion of this region Potzger distinguishes primarily between the pine forests and the deciduous or hardwood forests and discusses the difference between them (1946: 215-16):

> More species enter into the sociology of the pine associations than into that of the deciduous forest. . . . In most instances the pine associations have double the number of woody species of that found in the deciduous forest. It is very probable that light plays a prominent role in determination of this difference. The crown cover of the pine stands is open, perhaps only 85 percent coverage, while the crowns of the trees in the deciduous forest form a closed canopy. The greater penetration of light evidently enables even Quercus to invade the pine forest.
>
> . . . .
>
> Under the crown cover of the hardwoods there are really no shrub and small tree strata developed. The only shrub present at all regularly is Corylus . . . Quercus is not a component of the hardwoods complex. . . . Two transition tendencies are obvious in the hardwoods; one, which occurs where there is increasing soil moisture, is toward the cedar—spruce or fir—spruce forest of bogs; the other which occurs where there is decreasing soil moisture, is toward the pine association, especially toward the *Pinus strobus* variant. The hardwoods too, consist of two variants in which one or the other of the two dominants plays a more

important role. On upland sites, with decreasing soil moisture, *Acer saccharum* predominates; in the more moist lowland habitats *Tsuga canadensis* predominates. . . .

Most of the pine forest occurs in areas of rather short frostless seasons where agriculture was impractical. For this reason and because of the scarcity of natural food resources, Indian occupation of these forests was rather sparse except in localities where fishing was good. The spruce-fir and other forests of bogs, swamps, and highlands also were not much inhabited. Thus, the Indian population was concentrated mostly in the oak--hickory forests and in the birch--beech--maple--hemlock forests.

Hinsdale (1932: 3, 8) observed that "the Indian had a decided preference for the deciduous forests" except in special situations. "Where maple, beech, and oak predominated, the soil was more fertile and consequently afforded a richer growth of forage for animal life." This would seem to be less significant for subsistence, however, than the differences in agricultural potential. This is especially true when one considers that Newaygo County, Michigan, "which has the highest count of aboriginal sites of all the counties, was half covered with coniferous trees . . ." (Hinsdale 1932: 3). It may be added that Newaygo County has a relatively long average frost-free season.

The oak--hickory forests furnished acorns and nuts to Indian subsistence, while the beech--maple forests furnished beechnuts and maple sap; but there were no such important foods to be obtained in the coniferous forests unless they were burned out to be replaced temporarily by berry bushes. Of course, much of the territory was occupied by forests composed of a mixture of coniferous and deciduous trees; and most Indian habitations were along rivers and lakes where much of the forest probably had not developed to the climax.

Tree species that are prominent in the area but which were not specifically mentioned in the preceding discussion of vegetation types are as follows. Occurring predominantly in the south are rock elm, blackgum, sassafras, swamp white oak, black oak, box elder, red cedar, hackberry, black willow, peach-leaved willow, sycamore, shagbark hickory, bitternut hickory, butternut, cottonwood, white oak, and silver maple. Occurring throughout all, or most, of the area are slippery elm, blue beech, ironwood, bur oak, red oak, white elm, big-toothed aspen, quaking aspen, and balsam poplar. Generally these lists read from more southerly to more northerly species. For the most part, the conifers do not grow in the south, and the nut trees do not grow in the north. The maples, birches, elms, ashes, and poplars (including aspen and cottonwood) have rather wide distributions in eastern North America. This is true also of the plums, cherries, hazelnuts, serviceberries, hawthorns, dogwoods, sumacs, willows, viburnums, elders, and alders. Locust, coffee tree, red bud, paw-paw, hop tree, apple, and mulberry are primarily trees of southeastern and south-central United States which occur in the Upper Great Lakes region only in the southernmost portions.

Important Vertebrate Fauna

According to Dice (1938: 509-512) the following species of mammals range over most, or all, of the Canadian Biotic Province as he has defined it: black bear, raccoon, woodchuck, chipmunk, red squirrel, beaver, muskrat, porcupine, snowshoe hare, white-tailed deer, moose, and others. The black bear, red squirrel, beaver, and muskrat extend to the north and south of the Canadian province. The caribou ranged south into the northern part of the province, while the moose, snowshoe hare, and perhaps the porcupine had their southern limits near its southern border. The northern limits of the raccoon, chipmunk, and deer were near the northern boundary of the Canadian province, and the gray squirrel, elk, and bison ranged into the southern portion. The population and variety of mammals in the pine forests is very sparse.

The fish available to the Indians for food in the Upper Great Lakes region are sturgeon, pike, pickerel, lake trout, muskellunge, bass, bullhead, gar, greyling, sucker, sheepshead, and others. Also turtles, frogs, and clams were available as well as many birds including passenger pigeons, turkeys, grouse, and various water birds and their eggs (Hinsdale, 1932: 15-17).

Postglacial Vegetation History

It is likely that one or more of the vegetation types presently occupying the Great Lakes region have predominated throughout the period of human occupation but with different geographical distributions. Presumably, a narrow strip of treeless vegetation was present for relatively brief periods immediately following withdrawals of glacial ice, but the pollen diagrams for the region invariably begin with a representation of forest vegetation.

Sears (1942) presents the sequence for the Northeast as follows: a moist, cool period (maximum of fir and spruce); a dry, probably warmer, period (maximum of pine, often with oak); a more humid, also warm, period (maximum of beech and, in places, of hemlock); a warm, dry period (maximum of oak-hickory, minimum of beech); and the present, probably cooler and with more available moisture. He places the order of appearance of five forest genera in the northeastern part of the region as oak, hemlock, beech, hickory, and basswood, and in the southwestern part as oak, hickory, basswood, hemlock, and beech. With the exception of oak, these trees seem not to have become prominent in the area until after the pine maximum. This, however, could be due more to soil conditions than to climatic factors.

Pollen diagrams from bogs in Lower Michigan begin with a very high spruce pollen percentage which rapidly declines until it is almost entirely replaced by a high pine pollen percentage. The pine in turn declines, to be replaced by oak, except in the north where pine remains relatively high after an initial decline (after profiles published in Potzger, 1946: 240-41).

The initial high spruce percentage apparently represents an immediately postglacial spruce-fir forest which predominated until about 8000 B.C. but was thereafter replaced largely by pine, and to a lesser extent by oak, birch, willow, tamarack, elm, and probably other trees, followed by maple, beech, basswood, hemlock, and others. The pines reached a maximum about 6000 B.C. but later declined through an oak-pine period until the maximum northward spread of deciduous forest at about 2000 B.C. After that time the vegetation may have changed to become slightly more boreal in nature and developed into the modern vegetation patterns. These sequences were successively later and more condensed farther north (Griffin, 1961*a*; Zumberge and Potzger, 1956; Cain and Slater, 1948; and Potzger, 1946).

Cain and Slater (1948) have published a pollen diagram resulting from their studies of borings from Sodon Lake, Oakland County, Michigan. They found an early spruce domination replaced by pine which was in turn replaced by oak during a period of relatively high mesophytic hardwood pollen. Oak then gradually increases its dominance throughout the diagram without an evident indication of a xerothermic period.

Pine pollen in the Sodon Lake profile reaches its peak at the 23-foot level and drops considerably at the 22-foot level.

Samples of lake deposit from these two levels were combined to get a radiocarbon date of 5050 ± 200 B.C. (Crane, 1956: 668). This should date the early part of the period of replacement of pine pollen by oak pollen in Oakland County. Samples of peat representing the pine period at Jackson Camp and Cedar Bog Lake, Minnesota, have been dated at 5180 ± 300 B.C. and 6040 ± 420 B.C., respectively (Libby, 1952: 87). These dates agree remarkably well if it be assumed that a pine maximum existed in Minnesota at a slightly later time than it did in southern Michigan.

There are a number of wood and charcoal specimens from the Upper Great Lakes region which have been dated by radiocarbon analysis. Fortunately the trees they represent have been identified in many cases, by Stephen B. Preston, Volney H. Jones, myself, and others. As these increase in number they may be of greater help in determining postglacial history. Table II is a list of those which seem to be significant. Dates falling between 11,000 B.C. and 8000 B.C. are all of spruce, the later dates being from more northerly locations. Tamarack, white elm, and hemlock intervene before the first dated pine at 5300 B.C. and the first dated oak at 3350 B.C.; but oak and pine predominate in the list from then on. It is somewhat surprising that the correlation with the pollen sequence is so close in so short a list.

Prehistory and Paleoecology[1]

Man has occupied parts of the Upper Great Lakes region for perhaps 12,000 years and most, or all of it, for the past 9000. The first 2000 to 3000 years of this occupation was by people who apparently subsisted by means of hunting with spears tipped with fluted blades of flint. Mastodon, caribou, and other animals were available in the late glacial boreal forest environment. At present there is little evidence of the utilization of plant products during this period, but one can guess that wooden tools and weapons and perhaps cordage of plant fibers were made. It is likely that raspberry, cranberry, crowberry, and bearberry, and perhaps other plant foods that are available in northern Canada today were eaten (Table XI, p. 80).

From about 8000 to 3000 B.C. a warming trend resulted in the gradual establishment of the modern northern hardwood forest

[1] The chronology and archaeological assemblages discussed in this section are mainly according to Griffin (1961).

TABLE II

RADIOCARBON-DATED WOOD AND CHARCOAL FROM THE UPPER GREAT LAKES REGION[1]

Spruce	10,650 ± 300 B.C.	Steuben Co., Ind.	Crane, 1956: 669	M-350
Black spruce.	10,710 ± 400 B.C.	Lewiston, N.Y.	Rubin and Alexander, 1960: 135	W-861
Black spruce.	10,130 ± 300 B.C.	Lewiston, N.Y.	Rubin and Berthold, 1961: 86	W-883
Spruce	10,386 ± 350 B.C.	Sanilac Co., Mich.	Crane and Griffin, 1958: 1100	M-300
Spruce	8,900 ± 410 B.C.	Appleton, Wis.	Libby, 1954*a*: 139	C-800
Black spruce.	8,450 ± 300 B.C.	Two Creeks, Wis.	Crane, 1956: 669	M-343
Spruce	8,270 ± 250 B.C.	Ontonagon Co., Mich.	Crane, 1956: 669	M-359
Tamarack . . .	7,620 ± 500 B.C.	Lenawee Co., Mich.	Crane, 1956: 668	M-282
Tamarack . . .	5,510 ± 200 B.C.	Ingham Co., Mich.	Crane and Griffin, 1959: 175	M-638
White elm . . .	5,520 ± 250 B.C.	Isabella Co., Mich.	Crane, 1956: 669	M-360
Hemlock	5,320 ± 225 B.C.	Sanilac Co., Mich.	Crane and Griffin, 1958: 1100	M-299
Pine	5,300 ± 150 B.C.	Presque Isle Co., Mich.	Crane and Griffin, 1961: 106	M-1012
White elm . . .	3,920 ± 200 B.C.	Genesee Co., Mich.	Crane, 1956: 668	M-294
Hemlock	3,700 ± 150 B.C.	Mackinaw City, Mich.	Crane and Griffin, 1961: 106	M-855A
White oak . . .	3,350 ± 150 B.C.	Saginaw Co., Mich.	Crane and Griffin, 1960: 34	M-941
White cedar .	3,330 ± 150 B.C.	Saginaw Co., Mich.	Crane and Griffin, 1961: 105	M-1138
White oak . . .	2,930 ± 190 B.C.	Marinette, Wis.	Preston, *et al*, 1955: 958	Y-238
White pine . .	3,470 ± 200 B.C.	Michillinda, Mich.	Crane and Griffin, 1959: 174	M-474
White pine . .	3,200 ± 150 B.C.	Michillinda, Mich.	Crane and Griffin, 1959: 174	M-468
White oak . . .	2,940 ± 150 B.C.	Michillinda, Mich.	Crane and Griffin, 1959: 174	M-470
Oak	2,900 ± 150 B.C.	Michillinda, Mich.	Crane and Griffin, 1959: 174	M-471
White pine . .	2,850 ± 200 B.C.	Michillinda, Mich.	Crane and Griffin, 1959: 174	M-473
White elm. . .	2,800 ± 150 B.C.	Michillinda, Mich.	Crane and Griffin, 1959: 174	M-469
White pine . .	2,150 ± 125 B.C.	Michillinda, Mich.	Crane and Griffin, 1959: 174	M-472
White oak. . .	2,700 ± 200 B.C.	Blackwell, Ont.	McCallum and Dyck, 1960: 73	S-24
White oak. . .	2,650 ± 210 B.C.	Bright Grove, Ont.	McCallum and Dyck, 1960: 74	S-25
Red oak	2,130 ± 100 B.C.	Monroe Co., Mich.	Crane and Griffin, 1962: 183	M-1149
Oak, maple . .	1,980 ± 150 B.C.	Saginaw Co., Mich.	Crane and Griffin, 1962: 186	M-1139

[1] 1950 years subtracted to get B.C. dates as recommended by Flint and Deevey, 1962.

in the region as lake levels fluctuated in response to various geological activities. It probably was during this period that most of the economic plants became abundant enough for profitable utilization. This was likely a period of continuing adjustment to nature, especially as patterns of exploitation of the biota were involved. Hunting and fishing probably dominated subsistence pursuits, at least until nuts and tubers became easily available for food, and perhaps until the advent of agriculture, depending upon local conditions and season (Chapter III). It is likely that as population increased greater disturbance of the plant habitats took place, enabling certain food plants to increase in abundance (Chapter IV).

The archaeological assemblages of this period are generally referred to as Archaic. Like the Fluted Blade cultures they are not well known and are represented in this area mainly by a few scattered sites with rather scanty remains. This would indicate that population density remained at a low level and that the means of subsistence were not easily available with the technological inventory at hand. Representative sides of this period are Holcombe, Sheguiandah, George Lake, and various sites bearing remains of the Argillite Industry.

The Late Archaic in the Upper Great Lakes region lasted from about 3000 to 1000 B.C. and is represented by rather rich and varied tool and weapon assemblages which include adzes, gouges, grooved axes, awls, celts, knives, needles, punches, pikes, drills, chisels, wedges, bone fishhooks, barbed bone points, net and line sinkers, milling stones, pestles, and many projectile points in a variety of styles. Many of the artifacts are of copper and some are of ground stone.

The presence of woodworking tools indicates that the technological inventory included various implements of wood, perhaps including canoes. The possession of efficient axes would facilitate the collection of firewood and make tree sap and cambium more readily available during early spring when food sources would be hardest to find. This would raise the lowest point of seasonal food availability which would, in turn, allow the population to increase. It is possible that maple sap was first used in quantity at this time. Storage facilities also should have increased as a result of the possession of a wider variety of implements, which would tend to alleviate starvation during periods of low food availability and to even out what previously may have been more of a feast and famine subsistence cycle.

The presence of pestles and milling stones indicates that new plant foods were added to the diet or, at least, that there were more efficient means of preparation, especially of nuts and perhaps of certain seeds, dried tubers and berries, and cambium.

Artifact assemblages of the Late Archaic in the Upper Great Lakes region are referable to various cultures including Laurentian and Lamoka in the east and Old Copper and Glacial Kame in the west. This is the period during which the Feeheley site of Saginaw County, Michigan, was occupied, the plant remains from which are discussed in the following chapter.

The Early Woodland period follows the Late Archaic and persists until about 300 B.C. Burial goods are rather common and include copper celts and adzes and cache blades of flint that are triangular in form or pointed ovate with basal notching ("turkey tails"). Pottery appears at this time in the form of simple jars. Thus, for the first time, the Indians could boil down maple sap into storable sugar in significant quantity. It is not known, however, that this practice was extensive anywhere in pre-European times, and its time and place of origin are unknown. Apparently it was during this period that agriculture was first practiced in the Eastern Woodlands (Chapter V), but there is little or no evidence that it had as yet reached the Upper Great Lakes region.

The Middle Woodland period is represented in the region by cultures of Hopewellian affiliation, which seem to persist until about 500 A.D. They extended about half way up the west side of Lower Michigan and presumably were agricultural, being limited toward the north by short growing seasons and low temperatures (Chapter VI). Hopewellian sites have been located in Wisconsin, Michigan, and Ontario as well as in Illinois, Ohio, and other more southerly states. There are indications that this northern Hopewell penetration took place during a mild climatic phase (Griffin, 1958). "There are changes in the decorative styles and shapes of the pottery, of projectile points and other artifacts and indications of greater population density" (Griffin, 1961*a*: 153). Burial mounds were constructed, and the culture of the southern part of the Upper Great Lakes region generally takes on the aspect of a watered-down version of the spectacular Hopewell development farther south.

By 800 A.D. the Late Woodland cultures had made their appearance, and pottery styles had changed with cord impression coming into vogue. The Effigy Mound culture was present in Wisconsin from about 700 to 800 A.D. until perhaps 1400 A.D. Since it did not spread into those areas of northern Wisconsin, having the short-growing season, it is likely that subsistence was based to some extent on agriculture; however, there is not direct evidence of this.

By about A.D. 1000 there was influence from the Mississippian centers to the southwest. The southeastern part of the region was strongly influenced by the Owasco--Iroquoian development in the Lower Great Lakes region. There is abundant evidence that these cultures were agricultural (Chapter V).

Perhaps the earliest direct evidence of agriculture in the Upper Great Lakes region is from the Younge site in southeastern Michigan, where corn was found by Greenman (1937*b*). Corn from the Mackinac Straits area may be a slightly later type. In any case it would appear that the spread of agriculture in the Upper Lakes region was near its northernmost aboriginal limit by A.D. 1100 to 1200. There is evidence also that sunflowers, squash, and beans were cultivated during the latter part of the Late Woodland period; but there is no direct evidence of prehistoric agriculture north of the Straits of Mackinac.

A variety of cultures existed in the Upper Lakes region during the last few centuries preceeding European contact, according to Quimby (1960: 88-107). These include the Peninsular Woodland culture around the northern end of Lake Michigan, the Owasco culture in southeastern Michigan and adjacent Ontario, the Lalonde culture around Georgian Bay, the Fisher culture around the south end of Lake Michigan, the Lake Winnebago culture in eastern Wisconsin, and the Blue Island culture in the Chicago area. Presumably all of these cultures were agricultural to some extent.

Ethnogeography and Ethnohistory

The Indian tribes located around the Upper Great Lakes in early historic times include members of the Iroquoian, Siouan, and Algonkian linguistic families. The Iroquoians are represented by Huron, Tobacco (Petun, Tionontati), and Neutral. The only Siouan representative is Winnebago, though there may have been others during very early historic times. The Algonkians include Ottawa, Ojibwa (Chippewa), Menomini, Potawatomi, Mascouten, Miami, Sauk-Fox, and perhaps Kickapoo.

At the time of contact the Upper Lakes Iroquoians were located in Canada between Lake Huron and Lake Erie and in southeastern Michigan, and the Winnebago lived along the west side of Lake Michigan south of Green Bay. The rest of the region was occupied by Algonkian peoples. The authorities who were consulted place the Potawatomi and Mascouten in Lower Michigan and the Menomini in Upper Michigan when Europeans first reached the Upper Great Lakes in 1615. It is generally agreed that the Miami were located around the south end of Lake Michigan and that the Sauk, Fox, and Kickapoo were in Wisconsin during the early historic period (Kroeber, 1939; Kinietz, 1940; Driver and others, 1953; Murdock, 1960).

Since the Iowa, Oto, and Missouri speak dialects that are closely related to Winnebago, it is unlikely that they had been long separated from the Winnebago at the time of contact. The archaeological evidence would seem to support this (Griffin, 1960*b*). The Winnebago are separated from their sister tribes by the Sauk-Fox, and Kickapoo, who may not have occupied their locations in Wisconsin very long prior to contact.

Kroeber; Driver and others; and Murdock all locate the Ojibwa around most of Lake Superior from Sault Ste. Marie counter-clockwise to the vicinity of the Keweenaw Peninsula. The western and southern parts of this territory, however, were not occupied by them until rather late. According to Hickerson (1962: 2), the vast range of the Ojibwa ". . . had been occupied as a result of a series of migrations and conquests beginning in the ninth decade of the 17th century, originating in a rather small area adjacent to northern Lake Huron and eastern Lake Superior of which the great fishery at Sault Ste. Marie was the center." Thus, it is likely that Siouan speakers occupied the region around the west end of Lake Superior before European contact. The Ottawa are placed around the northern side of Georgian Bay and Manitoulin Island by most sources, though Murdock (1960) extends their range south and west into Michigan.

Though there may have been earlier influences (White, 1961), Jacques Cartier is the first European known to have approached the Lakes region. On his second voyage in 1534 he sailed up the St. Lawrence River to the Huron Villages of Stadacona and Hochelaga, where Quebec and Montreal now stand. "The loss of most of the manuscript of the third voyage, made in 1541, prevents certainty of anything more than the fact that the towns were still there at that time. When Champlain came up the river in 1603 they were gone, and Algonquin tribes were in possession and still advancing, having built forts up the River Richelieu" (Hunt, 1940: 13).

In 1615, on his third expedition to America, Champlain traveled inland to the country of the Hurons, between Lake Ontario and Georgian Bay. This visit was apparently the first by a European to the region of the Upper Great Lakes. At this time Champlain learned of the "Nation of Fire" who lived on the banks of Lake Huron and were at war with Indians he met (Deale, 1958: 305). Though the "Fire Nation" is often interpreted as the Potawatomi, it is likely that it included also some, or all, of the Mascouten and that these people were located in the Lower Peninsula of Michigan.

According to Hunt (1940: 109) the Potawatomi were the first to leave, retiring northward across the Straits of Mackinac and then along the northwestern shore of Lake Michigan to Wisconsin. The exodus apparently took place before 1634, because Nicolet includes the Potawatomi in the list of tribes around Green Bay at that time (Deale, 1958: 307). Hunt (1940; 112-113) also indicates that the Sauk and Fox lived in lower Michigan early in the seventeenth century, but Kinietz (1940), Quimby (1960), and Greenman (1961) do not find them there. Kinietz (1940: 309) states, however, that: "The next year, 1671, Dablon listed a few Winnebago living on Green Bay as in their own country, whereas the Potawatomi, Sauk, and the nation of the Fork were living there as foreigners, having been driven by fear of the Iroquois from their own country, the Lower Peninsula of Michigan. It is most likely that it was not the true Iroquois that drove them out, but the Neutrals." On the other hand, Greenman (1961: 14) states that the identification of the "people of the place of the fire" as Potawatomi ". . . is probably correct, but the supposition that the Potawatomi in 1616 were in the lower peninsula of Michigan appears to be incorrect. They must have been in eastern Wisconsin where . . . they had been seen in 1634 by Jean Nicolet. . . ."

II

PALEOETHNOBOTANY

Plant remains from archaeological sites in the Upper Great Lakes region are very scanty other than those from sites excavated by the University of Michigan Museum of Anthropology during the field seasons of 1960 to 1962. This field work, and the laboratory work carried out for the purpose of analyzing the materials recovered, was supported by a grant from the National Science Foundation. I identified and analyzed the plant remains with the generous assistance and advice of Volney H. Jones and Edward G. Voss.

One of the excavations took place at the Juntunen site (20 MK 1), a Late Woodland occupation site on Bois Blanc Island, Michigan, near the Straits of Mackinac. It was excavated under the direction of Alan L. McPherron and Albert T. Steegman during both seasons. The Feeheley site (20 SA 128) was excavated under the direction of David W. Taggart during the summer of 1960. This late Archaic site located near Saginaw, Michigan, was much less prolific of plant remains and other materials than was the Juntunen site. Analyses of these remains follow a compilation and discussion of archaeological plant food remains from the Midwest in general.

Plant Remains from Archaeological Sites in the Midwest

A survey was made in order to find records of archaeological remains of native food plants in the Upper Great Lakes region. This survey may have omitted some pertinent sources, but it appears that the only published references to some remains in the region are Quimby's (1939) records of pond nuts and hazelnut from Isle Royale and Wintemberg's (1946) records of plum, butternut, and acorn from Sidey--Mackay in Huronia. In an effort to obtain a meaningful sample the area considered was expanded to include Wisconsin, northeastern Illinois, northern Indiana, Ohio, northwestern Pennsylvania, western New York, and Ontario. This results in the addition of a number of sites in Ontario, New York, and Ohio and one each in Pennsylvania and Illinois, but none in Indiana and Wisconsin. Thus the distribution of the records presented here is somewhat better for the region of the Lower Great Lakes than for that of the Upper Great Lakes.[2]

The three records of native plant food remains for the Archaic period within the region outlined above are from the Frontenac Island site in Lake Cayuga, New York, with butternut (*Juglans cinerea* L.), hickory nut (*Carya* sp.), and acorn (*Quercus* sp.); from the Feeheley site near Saginaw, Michigan, with the remains of acorns, butternuts, hickory nuts, walnut (*Juglans nigra* L.), and grape (*Vitis* sp.); and from the Holcombe site (20 MB 30) with one charred chestnut (*Castanea dentata* [Marsh.] Borkh). Holcombe is an early Archaic site in Macomb County, Michigan, for which Griffin (1961*a*: 149) places the Archaic occupation at about 7000 B.C.; but there is also a Late Woodland occupation at this site. This is the only archaeological record of chestnut for Michigan. Charcoal from the site indicates that beechnuts (*Fagus grandifolia* Ehrh.), acorns, hickory nuts, and maple sap (*Acer* spp.) were available. Table III shows the analysis of the Holcombe charcoal, but it is not certain how much of it is representative of the Archaic occupation.

Early Woodland is represented by hickory nuts and acorns from the Vine Valley site in New York and by hickory nuts, walnuts, chenopod seeds (*Chenopodium* sp.), and blackberry or raspberry seeds (*Rubus* sp.) from the Florence, Cowan Creek, Goldsmith, Cordray, Dominion, Wolford, and Dunlap Adena sites in Ohio. All Adena sites are treated here as Early Woodland, though some radiocarbon dates indicate Middle Woodland period.

Five Middle Woodland sites were found to have produced the remains of native plant foods including hickory nut, walnut, acorn, plum, and hackberry at the McGraw site in Ross County, Ohio; hickory nut, walnut, and butternut at the Muntz (Sugar Run) site in Warren County, Pennsylvania; Canada plum (*Prunus nigra* Ait.) at the Serpent Mound in Ontario; butternut, walnut, and acorn at the Spoonville Hopewell site (Goodall Focus) in Ottawa County, Michigan; and acorn and walnut at the Schultz site (20 SA 2) in Saginaw County, Michigan. Thus there are at least 16 sites in the Midwest with pre-Late Woodland native plant food remains including 11 with hickory nut, 6 each with acorn and walnut, 4 with butternut, 2 each with chenopod seeds and plum, and 1 each with chestnut, blackberry, grape, and hackberry (*Celtis* sp.).

[2] The sources for these data are cited in the section "Native Food Plants by Season." Most of the collections from Ohio have been reported by Goslin (1952 and 1957). I identified most of those from Michigan and they are reported here for the first time.

TABLE III

ANALYSIS OF CHARCOAL FROM THE HOLCOMBE SITE (20 MB 30)

Charcoal	Occurrences				Percentages			
	Total	Plow Zone	c.9"-c.16"	c.16"-c.20"	Total	Plow Zone	c.9"-c.16"	c.16"-c.20"
Beech	66	10	51	5	34.2	27.0	38.3	21.7
Pine	35	4	27	4	18.1	10.8	20.3	17.4
Oak	31	6	18	7	16.1	16.2	13.5	30.4
red group	(16)	(1)	(10)	(5)	(8.3)	(2.7)	(7.5)	(21.7)
white group	(15)	(5)	(8)	(2)	(7.8)	(13.5)	(6.0)	(8.7)
Maple	24	4	14	6	12.4	10.8	10.5	26.1
Sycamore	12	4	8	..	6.2	10.8	6.0	...
Tulip tree	6	1	5	..	3.1	2.7	3.8	...
Birch	4	2	2	..	2.1	5.4	1.5	...
Chestnut	3	2	1	(nut)	1.6	5.4	0.75	...
Rock elm	3	2	1	..	1.6	5.4	0.75	...
Ash	2	1	..	1	1.0	2.7	...	4.3
Poplar	1	1	..	..	0.5	2.7	...	...
Blue beech	1	..	1	..	0.5	..	0.75	...
Basswood	1	..	1	..	0.5	..	0.75	...
Hickory	1	..	1	..	0.5	..	0.75	...
Black gum	1	..	1	..	0.5	..	0.75	...
Willow	1	..	1	..	0.5	..	0.75	...
Tamarack	1	..	1	..	0.5	..	0.75	...
Total	193	37	133	23	99.9	99.9	99.9	99.9

The Ohio cave sites include cultural remains from Late Woodland as well as from Middle Woodland complexes (Griffin, 1949: 83). From Ash Cave, Canter's Caves, and Kettle Hill Cave have come the remains of hickory nut, walnut, butternut, hazelnut (*Corylus* sp.), chestnut, acorn, grape, and chenopod. Thus the only addition is hazelnut which was found also at the Boyer's Run Rock Shelter (36 VE 6) in Pennsylvania along with hickory nuts (Thompson, 1954: 118).

Turning to the Late Woodland Period cultures of the region, 6 Fort Ancient sites have yielded the remains of 11 plant foods. Hickory nuts occurred at all 6 sites, plum pits (*Prunus americana* Marsh.) at 5, paw-paw seeds (*Asimina triloba* [L.] Dunal) at 4, butternut, walnut, and hazelnut (*Corylus americana* Walt.) each at 3, chestnut at 2, and acorn, grape, hackberry, and blackberry at 1 site each.

The remains of 5 native plant foods were found in 6 Owasco sites: Bates, Castle Creek, and Owasco Lake in New York, and Krieger, Miller, and Goessens in Ontario. Hickory nut and butternut occurred 3 times, acorn 2, and chestnut, grape, and chenopod once each. Chestnut and acorn are confined to the New York sites, while grape and chenopod are confined to the Ontario sites.

Thirteen sites of the Iroquoian Aspect in New York (Ripley, Westfield, Richmond Mills, Alhart, and Silverheels), Ontario (Middleport, Alway, Dutton, Lawson, Roebuck, Sidey-Mackay, "Aylmer," and Uren) and 2 sites in northern Ohio (Franks and South Park of the Whittlesey Focus) have produced the remains of 10 plant foods. Acorns are prominent for the first time, occurring in 8 sites. Butternut occurred in 8 sites also, but hickory nut appeared in only 5. Plum pits were found in 9 sites (*Prunus canadensis* in at least 3 cases and *P. americana* in at least two). Walnuts and cherry pits (*Prunus* spp.) occurred twice each, and beechnut, chestnut, and hazelnut occurred once each. Cherry, beechnut, and hazelnut were found only in Ontario, and the chestnut occurrence was in New York. Cherry and beechnut have not been mentioned up to this time.

Two sites which do not fall into any of the above categories are Fisher in northeastern Illinois and Marion in central Ohio. The former produced acorns, and the latter produced hickory nuts and plum pits.

The remaining sites are all in Michigan. Four of these are located in the southeastern part of the state. Hickory nuts were found at 20 MR 5 in Monroe County and at 20 MB 33 in Macomb County. Butternuts were found at the Norton site in Oakland County.

At Verchave II (20 MB 181) in Macomb County, a party from the University of Michigan Museum of Anthropology (summer, 1962) excavated a refuse pit containing small amounts of corn, acorn, grape seed, thornapple (*Crataegus* sp.), and walnut and a large quantity of broken hickory nutshells. All of this material had been completely carbonized.[3] The only time of year when all of the native plant remains could have been gathered is early October. Another pit yielded corn, grape seeds, plum pits, hazelnut, thornapple, and hickory nut. Again the season indicated is early October. Fishbones in the bottom of the first pit indicate a springtime occupation (Charles B. Cleland, personal communication, 1962). Thus, the site apparently was occupied in the spring (when plant food remains would be rare) and then again in early autumn.

In western Lower Michigan, acorn, walnut, and plum were represented at the Moccasin Bluff site in Berrien County; and grape seeds were found at the Dumaw Creek site in Oceana County. In Saginaw County part of a charred acorn was recovered from the Stroebel site (20 SA 14). The occurrences of pepperroot tubers (*Dentaria laciniata* Muhl.), chenopod seeds, hazelnuts, beechnuts, acorns, sumac seeds (*Rhus typhina* L.), grape seed, plum pits, cherry pits, bearberries (*Arctostaphylos uva-ursi* [L.] Spreng.), blueberries (*Vaccinium* sp.), blackberry seeds, elderberry seeds (*Sambucus pubens* Michx.), etc. at the Juntunen site will be discussed below.

Finally, three sites on Isle Royale have produced native plant food remains: pond nuts (*Nelumbo lutea* [Willd.] Pers.) at Chippewa Harbor, hazelnuts (*Corylus cornuta* Marsh.) at Indian Point, and bearberries at 20 IR 1/8.

Thus 21 different kinds of plant food remains from at least 25 species have been recovered from 59 archaeological sites in the Midwest for a total of 168 occurrences. The remains include 11 fruits and berries, 9 nuts, 1 small seed, and 1 tuber (ignoring the unidentified Juntunen tuber). Nuts account for 112 of the occurrences, fruits and berries account for 50, chenopod accounts for 6, and pepperroot for 1 . Of 109 occurrences that are definitely Late Woodland, 65 are nuts, 41 are fruits and berries, 2 are chenopod, and 1 is a tuber (Table IV).

These figures might indicate a greater relative dependence upon fruits and berries in Late Woodland times, but it probably indicates only differences in preservation and excavation technique. In any case it is unfortunate that archaeologists sometimes

[3] Since this paper was written charcoal from this pit, called Feature 2, has been radiocarbon dated as A.D. 1085 ± 100 (m-143) by the University of Michigan Radiocarbon Laboratory under the direction of Professor H. R. Crane. (J. B. G.)

have the attitude expressed by Meighan and others (1958: 133): "Plants, because of their perishable nature, are not usually present in archaeological sites." This might better read "Plant remains, because of their frequent small size and fragile nature, are not usually recovered during archaeological excavation" (Struever, 1962). It would appear that general analysis of archaeological native plant food remains is likely to have little significance until more have been collected, analyzed, and reported.

Several native plant food remains that might be expected to occur in Midwestern archaeological sites have not yet been recovered. These include groundnut tubers (*Apios americana* Medic.), Juneberry seeds (*Amelanchier* spp.), gooseberry seeds (*Ribes* spp.), *Viburnum* seeds, common elderberry seeds (*Sambucus canadensis* L.), strawberry achenes (*Fragaria virginiana* Duchesne), hackberry seeds, apple seeds (*Malus* spp.), and wild rice (*Zizania aquatica* L.).

There are many occurrences of cultigen remains (i.e., corn, beans, squash, gourd, sunflower) from archaeological sites in the Midwest. These will be discussed in Chapter V.

The Feeheley Site

The Feeheley site (20 SA 128) is a late Archaic burial site located in Saginaw County, Michigan, about five miles west of the junction of the Tittabawassee and Saginaw Rivers at the city of Saginaw. Radiocarbon analysis of charcoal that was found with an occupational feature produced a date of 1970 ± 150 B.C. (Crane and Griffin, 1962: 186).

The following description is from David W. Taggart's abstract of his unpublished report:

> The Feeheley site represents an early Point Peninsula and terminal Laurentian complex . . . with a burial complex very similar to that reported for Glacial Kame sites. The site is located upon a small sand knoll southwest of Saginaw which represents a sand dune formed during immediately post-Nippissing maximum times (after 3000 B.C.) along the shore of a low peninsula jutting into a considerably extended embayment of the present Saginaw Bay. . . . Occupational debris consists of a thin accumulation of fire-cracked rock, chert spalls and cores, projectile points, scrapers, knives, and hammerstones. . . . Occupational debris and features are spread over the site and the pits are interdigitated with the burials which show superposition in a small constricted area of the site. The burial pattern is typically one of tightly flexed burial within skin or bark bundles, with the inclusion of copper necklaces and celts or awls, shell ornaments, and at times food offerings of nuts, placed within small pits with a profusion of red ochre. . . .

TABLE IV

ARCHAEOLOGICAL NATIVE PLANT FOOD REMAINS IN THE MIDWEST

	Archaeological Sites	Pepperroot	Elderberry	Blackberry	Grape	Blueberry	Bearberry	Sumac Berry	Hackberry	Cherry	Plum	Thornapple	Paw-Paw	Hazelnut	Chestnut	Beechnut	Walnut	Butternut	Hickory Nut	Acorn	Pond Nut	Chenopod
Archaic	Holcombe-Mich.														x							
	Frontenac-N.Y.																	x	x	x		
	Feeheley-Mich.				x												x	x	x	x		
Early Woodland	Vine Valley-N.Y.																		x	x		
	Corcray-Ohio																		x			
	Dominion-Ohio																		x			
	Wolford-Ohio																		x			
	Dunlap-Ohio																		x			
	Goldsmith-Ohio			x																		
	Cowan Creek-Ohio																		x			x
	Florence-Ohio																x		x			x
Middle Woodland	McGraw-Ohio								x		x						x		x	x		
	Muntz-Pa.																x	x	x			
	Serpent Mound-Ont.										x											
	Spoonville-Mich.																x	x		x		
	Schultz-Mich.																x			x		
Caves	Ash-Ohio				x									x			x	x	x	x		x
	Kettle Hill-Ohio				x									x	x		x	x	x			x
	Canters-Ohio				x									x	x			x	x			
Owasco	Bates-N.Y.														x			x		x		
	Castle Cr.-N.Y.																	x	x	x		
	Owasco L.-N.Y.																		x			
	Krieger-Ont.				x																	
	Goessens-Ont.																		x			x
	Miller-Ont.																	x				

TABLE IV *(CONT.)*

	Archaeological Sites	Pepperroot	Elderberry	Blackberry	Grape	Blueberry	Bearberry	Sumac Berry	Hackberry	Cherry	Plum	Thornapple	Paw-Paw	Hazelnut	Chestnut	Beechnut	Walnut	Butternut	Hickory Nut	Acorn	Pond Nut	Chenopod
Fort Ancient	Baum—Ohio	. .	. .	. .	x	. .	. .	. .	. .	. .	x	. .	x	x	x	. .	x	x	x	. .	. .	. .
	Gartner—Ohio	. .	. .	. .	. .	. .	. .	. .	. .	. .	x	. .	x	x	x	. .	x	x	x	. .	. .	. .
	Feurt—Ohio	. .	. .	x	. .	. .	. .	. .	. .	. .	x	. .	x	. .	. .	. .	x	x	x	x	. .	. .
	Baldwin- Ohio	. .	. .	. .	. .	. .	. .	. .	. .	. .	x	. .	. .	. .	. .	. .	. .	. .	x	. .	. .	. .
	Cramer—Ohio	. .	. .	. .	. .	. .	. .	. .	x	. .	x	. .	x	x	. .	. .	. .	. .	x	. .	. .	. .
	Schizler—Ohio	. .	. .	. .	. .	. .	. .	. .	. .	. .	. .	. .	. .	. .	. .	. .	. .	. .	x	. .	. .	. .
Iroquoian	Ripley—N.Y.	. .	. .	. .	. .	. .	. .	. .	. .	. .	x	. .	. .	. .	. .	. .	. .	x	x	. .	. .	. .
	Westfield—N.Y.	. .	. .	. .	. .	. .	. .	. .	. .	. .	x	. .	. .	. .	x	. .	x	x	x	x	. .	. .
	Richmond Mills—N.Y.	. .	. .	. .	. .	. .	. .	. .	. .	. .	. .	. .	. .	. .	. .	. .	. .	x	x	x	. .	. .
	Alhart—N.Y.	. .	. .	. .	. .	. .	. .	. .	. .	. .	. .	. .	. .	. .	. .	. .	. .	. .	x	. .	. .	. .
	Silverheels—N.Y.	. .	. .	. .	. .	. .	. .	. .	. .	. .	. .	. .	. .	. .	. .	. .	. .	. .	. .	x	. .	. .
	Middleport—Ont.	. .	. .	. .	. .	. .	. .	. .	. .	. .	x	. .	. .	. .	. .	. .	. .	. .	. .	. .	. .	. .
	Alway—Ont.	. .	. .	. .	. .	. .	. .	. .	. .	x	x	. .	. .	. .	. .	. .	. .	. .	. .	x	. .	. .
	Dutton—Ont.	. .	. .	. .	. .	. .	. .	. .	. .	. .	x	. .	. .	. .	. .	. .	. .	. .	. .	. .	. .	. .
	Lawson—Ont.	. .	. .	. .	. .	. .	. .	. .	. .	. .	x	. .	. .	. .	. .	. .	. .	x	x	x	. .	. .
	Roebuck—Ont.	. .	. .	. .	. .	. .	. .	. .	. .	x	x	. .	. .	x	. .	x	. .	x	. .	. .	. .	. .
	Sidey-Mackay—Ont.	. .	. .	. .	. .	. .	. .	. .	. .	. .	x	. .	. .	. .	. .	. .	. .	x	. .	x	. .	. .
	Aylmer—Ont.	. .	. .	. .	. .	. .	. .	. .	. .	. .	. .	. .	. .	. .	. .	x	x	x	. .	x	. .	. .
	Uren—Ont.	. .	. .	. .	. .	. .	. .	. .	. .	. .	. .	. .	. .	. .	. .	. .	. .	x	. .	. .	. .	. .
	Franks—Ohio	. .	. .	. .	. .	. .	. .	. .	. .	. .	x	. .	. .	. .	. .	. .	. .	. .	. .	x	. .	. .
	South Park—Ohio	. .	. .	. .	. .	. .	. .	. .	. .	. .	. .	. .	. .	. .	. .	. .	. .	x	. .	. .	. .	. .

TABLE IV *(CONT.)*

Late Woodland	20 MR 5—Mich.	. .	. .	. .	. .	. .	. .	. .	. .	. .	. .	. .	. .	. .	. .	. .	. .	. .	x	. .	. .	. .
	Norton—Mich.	. .	. .	. .	. .	. .	. .	. .	. .	. .	. .	. .	. .	. .	. .	. .	. .	x	. .	. .	. .	. .
	20 MB 33—Mich.	. .	. .	. .	. .	. .	. .	. .	. .	. .	. .	. .	. .	. .	. .	. .	. .	. .	x	. .	. .	. .
	Verchave II—Mich.	. .	. .	. .	x	. .	. .	. .	. .	. .	x	x	. .	x	. .	. .	x	. .	x	x	. .	. .
	Moccasin Bluff—Mich	. .	. .	. .	. .	. .	. .	. .	. .	. .	x	. .	. .	. .	. .	. .	x	. .	. .	x	. .	. .
	Dumaw Cr.—Mich.	. .	. .	. .	x	. .	. .	. .	. .	. .	. .	. .	. .	. .	. .	. .	. .	. .	. .	. .	. .	. .
	Stroebel—Mich.	. .	. .	. .	. .	. .	. .	. .	. .	. .	. .	. .	. .	. .	. .	. .	. .	. .	. .	x	. .	. .
	Juntunen—Mich.	x	x	x	x	x	x	x	. .	x	x	. .	. .	x	. .	x	. .	. .	. .	x	. .	x
	Isle Royale—Mich.	. .	. .	. .	. .	. .	x	. .	. .	. .	. .	. .	. .	x	. .	. .	. .	. .	. .	. .	x	. .
Other	Fisher—Ill.	. .	. .	. .	. .	. .	. .	. .	. .	. .	. .	. .	. .	. .	. .	. .	. .	. .	. .	x	. .	. .
	Marion—Ohio.	. .	. .	. .	. .	. .	. .	. .	. .	. .	x	. .	. .	. .	. .	. .	. .	. .	x	. .	. .	. .
	Total	1	1	3	9	1	2	1	2	3	20	1	4	10	7	3	15	23	32	23	1	6

The food plant remains were found at about 20 locations within an area of about 100 feet by 300 feet and up to about 3.5 feet deep. Most of them were found in the area of most extensive excavation at the east end of the site. Butternuts were found at 11 locations, acorns at 6, hickory nuts at 4, and walnut and grape seed each at 1 location. Butternut occurred once each with walnut and hickory nut, and acorn apparently occurred once with hickory nut, but other occurrences were isolated. Charcoal present at the site indicates that beechnuts and maple sap were available also.

At the west end of the site butternut occurred once. Hickory nut occurred at 3 locations in 2 excavations, and butternut occurred at 4 locations in 3 excavations near the middle of the site. In the main excavation area 1 grape seed occurs, walnut and hickory nut each occur once, butternut occurs at 6 locations in 4 excavated squares, and acorn occurs at 6 locations in 6 excavated squares.

Butternut was found in one occupational feature. Hickory nut was found in 1 occupational feature with acorn. Acorn was found in 4 occupational features and with a burial. At least 13 acorns were represented in the burial, but they are the only food plant remains mentioned so far which were not carbonized. It is possible that they were intrusive, because there was a rodent burrow in one part of the burial. The acorn shells, however, are from another part of the burial; they are covered with red ochre, and they were located in the immediate vicinity of a copper celt.

The season of availability of hickory nut, walnut, and butternut is October. For acorn it is September to November, and for grape it is August to autumn. Thus it appears that the occupation or occupations of the site represented by the plant food remains took place in early autumn if not at other seasons.

Butternut was found as deep as 3.5 feet; acorn, walnut, and hickory nut to more than 2.0 feet deep; and the grape seed at about 0.6 of a foot deep.

The geographic range of walnut barely reaches north to Saginaw Bay at the present time. This would indicate why walnut was found only once in the site. It would also indicate that the climate at the time of occupation was no cooler than it is at the present time. Butternut and hickory range northward about two-thirds of the way up the Lower Peninsula at the present time.

Burials were commonly accompanied by necklaces of copper beads strung on cord which was well preserved. The cord was manufactured by twisting strands of fiber from the inner bark of a woody plant. It is quite likely that the fibers are those of the basswood tree (*Tilia americana* L.), but there are other possibilities.

A large number of small pieces of charcoal were excavated from the Feeheley site. I identified a majority of these. Some of the charcoal which was examined remains unidentified, however; so the following list of woods does not include all of those which were present.

A total of 10 genera and at least 13 species of trees are represented in the 35 lots of charcoal identified by the writer. Table V shows the results of these identifications. It is interesting that hickory occurred at only three locations and that walnut and butternut occurred not at all. This may mean that these trees were rather scarce or that they were usually spared from furnishing firewood. Most firewood, however, probably was picked up from the ground.

It is tempting to describe the vegetation of the region at the time of the Feeheley occupation as oak dominated but with large numbers of ash, pine, beech, and sycamore. There are, however, at least four reasonable objections to this. One is that of selective gathering of firewood. Another is the possible overrepresentation of certain trees by the deadwood on the ground. A third is is an overrepresentation of certain species because of strictly local environmental conditions which perhaps were primarily hydric at the time because of the high lake level. Finally, we have no assurance that all of the charcoal is contemporaneous.

Oaks grow in a variety of habitats, but most of those growing in the Saginaw area (primarily the red, white, and bur oaks) have a tendency to grow in moist to damp or even wet soils. The pine charcoal appears to be mostly from white pine which prefers moist sandy soils. The only large tree directly on the Feeheley site is a dead red pine. Beech, basswood, maples, white elm, and the white ash group grow in a variety of moist to wet soils. The hickories and slippery elm grow on various soils. Black ash, sycamore, and blue beech almost always grow in wet soils around swamps and streams. Thus the assemblage appears to be primarily one of bottomlands and swamp or stream borders with some drier locations for pine, beech, hickory, basswood, and perhaps oak, elm, and maple. The high percentage of black ash and sycamore definitely indicates rather wet areas nearby, but if these species and those with low representations be eliminated the picture is one of oak forest with considerable numbers of pine, beech, and perhaps ash present. Since beech and ash are not heavy pollen producers like oak and pine, such a forest would likely be represented in pollen studies as oak--pine.

Zumberg and Potzger (1956) have made geological and pollen studies of an excellent vertical exposure of buried bog and beach

TABLE V

ANALYSIS OF CHARCOAL FROM THE FEEHELEY SITE (20 SA 128)

Square: level (feature)	Red Oak Group	White Oak Group	Pine	Beech	Sycamore	Black Ash	Ash	White Ash Group	Hickory	Maple	Blue Beech	White Elm	Slippery Elm	Basswood
30-110:1.8'			x		x		x							
80-110:1.2	x													
:4-5							x							
90-110:(8)	x													
:(D)	x	x												
30-115:1	x										x			
40-115:0.7'-1	x	x	x											
60-120:(J)	x	x		x	x		x							x
70-120:(C)						x								
:(AE)	x	x	x			x		x						
70-140:(3)	x	x	x			x								
80-120:(E)	x	x						x				x	x	
90-120:(P)	x			x										
65-125:3-4		x		x	x									
60-130:2-3				x	x					x				
:(4)	x				x									
:(N)				x	x					x				
70-140:(L)	x	x	x			x		x	x					
100-140:1		x	x	x		x								
110-140:1			x	x										
60-150:(R)			x											
65-190:0-0.5'			x											
55-250:(AN)	x													
30-300:1-2'				x	x	x			x					
:2	x													
:(Y)	x	x	x	x			x				x	x		
10-305:1-2	x	x	x	x	x	x								
:1a			x											
5-315:	x	x		x			x							
5-324:1.4'	x	x		x										
40-350:0.7'	x		x											
:1'		x												
70-400:(M)	x	x	x		x	x	x							
100-445:(V)	x	x							x					
Total	22	17	14	12	9	8	6	3	3	2	2	2	1	1

sediments on the shore of Lake Michigan near South Haven, Michigan, which is at a latitude about 70 miles south of the Feeheley site. Several of the strata were dated by radiocarbon analysis, the youngest of which, at 2000 ± 175 B. C. represents a time that closely approximates the Nippissing maximum. Pollen analysis of this stratum indicates the presence of an oak—hemlock—broad-leaved forest which was immediately preceded and succeeded by oak—pine in slightly cooler (or dryer?) periods. Thus the indication of forest type by the Feeheley charcoal would appear to agree rather closely with the results of pollen analysis. The Lake Michigan shore ecology, however, may have been somewhat different from the Saginaw Bay shore ecology of that time.

On his map of presettlement forest in Michigan, Veatch (1959) indicates that the forest in the vicinity of the Feeheley site was composed of elm, silver maple, ash, swamp white oak, basswood, shagbark hickory, sycamore, cottonwood, red oak, and bur oak. The composition of the Feeheley charcoal sample disagrees with that of Veatch's list primarily in the large percentage of beech and pine in the former. This is difficult to interpret because of the relative divergence of the environmental preferences of these two trees. It is possible that some of the charcoal identified as pine is actually hemlock. Beech is not a cold-climate tree. It "reaches greatest size and best development farther south and, during the exceptionally cold winter of 1934, suffered considerable damage in the northeast" (Harlow, 1957: 135). On the other hand the Feeheley site is near the southern limits of the ranges of the Michigan pines. Red pine and jack pine prefer rather dry sandy soils. White pine prefers light, fertile loam and damp, sandy soils. Beech is "Found on a wide variety of soils but nearly always where abundant surface moisture is available" (Harlow, 1957: 135). It seems likely that diverse habitats were present in the area then as now.

Trees growing in the immediate vicinity of the Feeheley site at the present time are white pine, white oak, swamp white oak, red oak, elm, birch, beech, maple, basswood, aspen, red ash, black cherry, prickly ash, Juneberry, apple, and hawthorn.

The Juntunen Site

The Juntunen site is located at the northwestern tip of Bois Blanc Island and about 8 miles due east of the bridge connecting the Upper and Lower Peninsulas of Michigan at the Straits of Mackinac. The island is situated at the northwestern end of Lake Huron in Mackinac County. Its southern coastline parallels the coast of Cheboygan County in Lower Michigan about 4 miles to the south. Between 600 and 700 yards from the shore adjacent to the site is Round Island, just beyond which is Mackinac Island.

Bois Blanc Island is about 11.5 miles long and ranges in width from about 1 mile at the west end to about 4.5 miles at the middle and at the east end. It includes four lakes, each about 1 mile in its greatest dimension, and about thirty smaller lakes and ponds, most of which are in the eastern half of the island. There are many bogs and much of the land is swampy, but there are very few streams. The whole island is low with the maximum elevation above lake level probably not much greater than 60 feet.

The vegetation of Bois Blanc is varied and includes both coniferous and hardwood forests. A generalized forest cover map, made up in 1945 by B.C. Jenkins and C. H. Wonser for the Game Division of the Michigan Department of Conservation, shows four types of vegetation. Each of three forest types is well distributed throughout the island. They are "*spruce*--balsam--cedar," "*poplar*--spruce--balsam--cedar," and "*hardwood*--second growth." The fourth vegetation type, "*open*--farm--grass," consists of seven relatively small areas that are rather widely scattered. I observed, however, a number of grassy openings which were not indicated on the map.

There is one area on the island that is said to be in virgin forest. This is on and around a large point that projects into Lake Huron from the northeastern side of Bois Blanc. Here I found very few food plants. Only the western part of this forest was examined and it was found to be made up mostly of white and red pine, white cedar, yellow birch, and hard maple.

The largest area of hardwood forest is to be found inland at the west end of the island. My observation is that this forest is almost pure beech--maple with very few economic food plants other than beech and maple except around clearings and bogs. The surrounding coastal area here is spruce--balsam--cedar. It is in this latter type of forest that the Juntunen site is situated but with many hardwoods immediately surrounding the attenuated grassy clearing, within which the site is located on a raised beach about 15 feet above lake level.

At the edge of the clearing there are a number of plants which produce berries in the summer. They include bearberry, Juneberry, fire cherry, bunchberry, elderberry, and partridge berry; and the common milkweed grows abundantly within the clearing. Some other economic plants which grow upon the island more or less abundantly include paper birch, white cedar, sugar maple, beech, oak, viburnum, sumac, sand cherry, gooseberry, blackberry, chokecherry, two-leaved Solomon's seal, beach pea, strawberry, marsh marigold, and blueberry. Of course basswood, called *bois blanc* by the French, also grows on the island.

Unfortunately there is no published report of the flora of Bois Blanc Island, and as a result it is poorly known. There is archaeological evidence of four food plants which are not known by myself to occur there today, though they probably are present. These are hazel, Canada plum, grape, and pepperroot (toothwort).

Juntunen is a Late Woodland site with occupations beginning perhaps about 800 A.D. and lasting perhaps into the historic period. Three radiocarbon dates on excavated materials from deepest to shallowest in the same pit are A.D. 835 ± 75, A.D. 910 ± 75, and A.D. 1070 ± 75 (Crane and Griffin, 1961: 110). The occupation levels from which the dated materials were excavated are separated by sterile sand layers, and the earliest date represents the earliest known occupation in the excavation squares. A radiocarbon date of A.D. 1320 ± 75 was derived from materials excavated with an ossuary in another part of the site (Crane and Griffin, 1963: 233). Thus there would seem to have been a series of occupations which lasted for at least 500 years. The depth of materials from the earliest occupation at the location which furnished the dated material is about four feet. Most of the organic remains consist of fish bone and charcoal. The majority of the cultural remains are potsherds and flint chips.

The method of excavation of the Juntunen site was to dig more than 50 squares, most of which were 10 by 10 or 5 by 10 feet square. Plant remains (excluding wood charcoal and nonarchaeological remains) were recovered, examined, and mostly identified from 24 of these squares, 14 of which were 10 by 10 and 10 of which were 5 by 10 feet square. Eight of the 10 by 10 pits are contiguous so as to form a single irregular excavated area, but otherwise contiguous pits rarely make up excavated areas of more than 10 by 20 feet. Thus, with the exception of the larger excavated area, the plant remains are from rather scattered locations but all from within an area of about 150 by 300 feet which is well centered in the approximately 200 by 600 feet total area of the site.

The stratigraphy of the site is extremely complex. There are about 25 recognizable strata with lenses, pockets, pits, and various disturbances intermingled among them. Most of the strata contain evidences of Indian occupation but some of them are sterile

TABLE VI

STRATIFICATION AT THE JUNTUNEN SITE (20 MK 1) AND POSITION OF SOME PLANT REMAINS

Occupation Level	Square								
	Stratum	1600-1130	1605-1130	1700-1120	1700-1140	1710-1140	1710-1180	1870-1140	1880-1130
	25c								
	25b								
				0.8'	0.7'	0.7'		0.7'	
F					ACGL			ABCG	
(clearing)?). .	25a			BPQZ	PRS	CF		PRZ	
		0.9'		1.0'	0.9'	1.1'	0.8'	0.9'	
	24	PxZ					BZ		
		1.0'					1.0'		
E?	23			1.0'					
	22			BGPnZ					
				1.2'					
	21								
				1.3'					
E	20			BCPnZ					
				1.5'					
	19			BCT					
	18			VZ					
		1.1'		3.2'					1.0'
D									1070 A.D.
(clearing?) . .	17	BT		B					BC
		1.3'		4.5'					1.5'
	16								910 A.D.
									1.7'
									910 A.D.
									BC

TABLE VI *(CONT.)*

C	15								
(clearing?) . .									2.0'
C	14								
	13								
B (clearing?) . .	12								
									2.6'
	11								C
									3.7'
	10								
	9								
	8								
	7								
	6								
		1.6'						3.0'	
A									
(podsol)	5	B						C	835 A.D.
		1.7'						3.3'	
	4								
		1.8'	2.2'						
(podsol)	3b	B	C						
		2.0'	2.3'						
	3a								
	2								
(parent material) . .	1								

KEY:

A *Arctostaphylos* - bearberry
B *Betula* - birchbark
C *Corylus* - beaked hazelnut
F *Fagus* - beechnut
G *Galium* sp.
L *Chenopodium* - lambsquarter
P *Prunus* - fire cherry
Px *Prunus* - cherry
Pn *Prunus* - Canada plum
Q *Quercus* - acorn
R *Rubus* - blackberry, raspberry
S *Sambucus* - elderberry
T Tuber (unidentified)
V *Vaccinium* - blueberry, cranberry
Z *Zea* - corn

of artifactual materials. Extensive analysis of the confusing jumble of artificial and natural strata by Henry Wright led to his determination of six major levels of occupation signified by the letters A through F. Levels A, C, and D are the levels dated at A.D. 835, 910, and 1070, but level F apparently represents the major occupation, and it is this level from which a majority of the plant remains appear to be derived. It may be represented by the A.D. 1330 date (Table VI).

Excluding wood charcoal, I identified the remains of 18 different plants from archaeological contexts. Several specimens each of a tuber and what is apparently a large seed about one centimeter long were found but not identified. Most of the larger plant remains were found by low-power microscopic examination of samples of charcoal remaining after gross sifting in the field. Most of the smaller seeds were located by low-power microscopic examination of the mixed detritus that remained after fine sifting of cultural deposits in the laboratory. Altogether, about 80 lots of identifiable plant remains were located (excluding wood charcoal, etc.) and about 20 lots that included nothing which could be definitely identified or apparently consisted of intrusive materials. Identifications were accomplished by direct microscopic comparisons with check materials which in most cases were taken from living plants, but in many cases archaeological remains previously identified by Volney H. Jones were especially helpful.

Of the identified plant remains, 15 are apparently food materials, 2 are weed seeds, and 1 is birchbark which occurred here and there throughout the site. The food remains included kernels of corn, fire cherry pits, pits of one or more other cherries (sand cherry, black cherry, or chokecherry), pits of Canada plum, hazel nuts, bearberries, chenopod seeds (apparently mostly *Chenopodium album*, but one collection appears to be *C. hybridum* var. *gigantospermum* Rouleau), seeds of blackberry, raspberry, etc., blueberries or cranberries, grape seed, elderberry seeds, beechnut fragments, acorn fragments, sumac seeds (probably *Rhus typhina* or *R. glabra*), and pepperroot tubers. The weed seeds are of the genus Galium (bedstraw, cleavers), perhaps *G. triflorum* Michx. and a *Rumex* species (dock, sorrel), perhaps *R. verticillatus* L. or *R. orbiculatus* Gray. Altogether there were 153 significant finds of plant remains (Table VII).

Birchbark and hazelnut are each represented in two squares and at two levels in Occupation A, the earliest occupation at the site. Hazelnut, and probably also birchbark occur in occupation B. An unidentified cherry occurs in B or C; and birchbark and

TABLE VII

PLANT REMAINS FROM THE JUNTUNEN SITE (20 MK 1)

Plant Remains	Carbonization	Squares	Locations
Birchbark *(Betula papyrifera)*	N, C	13	27
Hazel nut *(Corylus rostrata)*	C	14	25
Corn *(Zea mays)*	C	11	16
Fire cherry *(Prunus pensylvanica)*[1] .	N, C	12	14
Blackberry, etc. *(Rubus* spp.)	N, C	8	10
Cleavers *(Galium* sp.)	C	7	7
Elderberry *(Sambucus pubens)*	N, C	6	7
Unidentified tuber	C	6	7
Unidentified "large seed"	C	5	7
Chenopod *(Chenopodium* spp.)	N, C	6	6
Bearberry *(Arctostaphylos uva-ursi)*.	C	4	4
Sumac berry *(Rhus typhina?)*	N, C	4	4
Beechnut *(Fagus grandifolia)*	C	4	4
Acorn *(Quercus* sp.)	C	4	4
Pepperroot *(Dentaria laciniata)*	C	2	3
Cherry *(Prunus* sp.)	C	2	3
Canada plum *(Prunus nigra)*	C	1	2
Grape *(Vitis* sp.)	C	1	1
Blueberry *(Vaccinium* sp.)	C	1	1
Dock *(Rumex* sp.)	C	1	1
Total		102	153

[1] Exclusive of those from rodent burrows.

Note: C = Carbonized; N = Noncarbonized.

hazelnut occur together in C, D, E, and F. The large tuber occurs in occupation D with birchbark; and it occurs above D, but possibly below E, with corn, hazelnut, blueberries, and birchbark. This latter occurrence is most likely attributable to occupation E which is later than the 1070 date. This is the only occurrence of blueberries, but it was only by rare chance that they were preserved in an identifiable condition. Identification was accomplished as a result of examining the minute seeds which would not have been found had the fragile cluster of berries not been preserved and recovered.

Occupation E probably represents the time of the earliest corn growing, though one instance of the occurrence of corn may be somewhat earlier. In any case, corn was as abundant as anything else in deposits identified with occupation E, where it occurs in at least two levels of one square with Canada plum, birchbark, and hazelnut and in a higher level in two other squares, in one with an unidentified cherry pit and in the other with birchbark. This higher level may actually be a part of occupation F. The large unidentified seed apparently occurs for the first time in occupation E. It was found with birchbark.

Everything else seems to occur for the first time in occupation F, and at the same level. This includes acorn, blackberry and/or raspberry, fire cherry, bearberry, beechnut, elderberry, chenopod, *Galium*, probably grape and sumac, and perhaps also the pepperroot tubers and *Rumex*. Those plant remains which definitely were at this level occurred together in three sets as follows: (1) corn, fire cherry, acorn, and birchbark; (2) corn, fire cherry, "blackberry," bearberry, elderberry, hazelnut, birchbark, and *Galium;* and (3) fire cherry, "blackberry," bearberry, hazelnut, beechnut, chenopod, and *Galium*. At one location, at a higher level, corn (the large seed), and birchbark occur (Table VI).

The locations as presented above account for fewer than 30 lots of plant remains out of the total of about 80 from which identifications were made. The remaining lots have not yet been placed in stratigraphic association with the occupation levels. It is in these latter lots that the grape seed, the two occurrences of the pepperroot tuber, and the four occurrences of sumac seeds occur. In every case, however, sumac seeds and the grape seed occur with or above that group of plants that is known to occur only at occupation level F. It is possible that the *Rumex* seed and the earliest occurrence of the pepperroot tubers is below occupation F.

The only plant remains identified from a depth of more than 1.5 feet are birchbark (to 4.5 feet deep), hazelnut, unidentified cherry, corn, blueberry, and the unidentified tuber. Of these, the deepest tuber is no earlier than occupation D, the blueberries are later than occupation D, and the corn (with one possible exception) is probably all later than occupation D, though not definitely so.

The *Rumex* seed occurs only with hazelnut and chenopod. The Canada plum and the large seed occur only with hazelnut, corn, and birchbark, though the latter occurs in four locations alone rather near the surface. The unidentified tuber occurs only with hazelnut, corn, birchbark, and blueberry, and in four locations alone. The blueberries occur only with hazelnut, corn, birchbark, and the tuber. Unidentified cherries occur with corn, pepperroot tubers, elderberry seeds, and perhaps also with fire cherry, blackberry, and birchbark. Birchbark, corn, and hazelnut occur most often together; and they occur with anything else except the grape seed with the following exceptions: only hazelnut occurs with the *Rumex* seed, only birchbark occurs with pepperroot tubers, and only corn definitely occurs with an unidentified cherry pit.

The rest of the identified plant remains occur mostly together, but also with corn, hazelnut, and birchbark, with the exceptions noted above. Though the pepperroot tuber does not occur with chenopod, acorn, grape, or *Galium*, it occurs twice each with elderberry, beechnut, and sumac seeds. The single occurrence of grape seed is with elderberry, fire cherry, blackberry, acorn, chenopod, and *Galium*. Elderberry, fire cherry, blackberry, chenopod, *Galium*, beechnut, sumac, acorn, and bearberry all occur with each other from 1 to 10 times each and with corn, hazelnut, and birchbark from 1 to 4 times each. Elderberry, fire cherry, blackberry, chenopod, and *Galium* occur with each other from 4 to 8 times each except for fire cherry and blackberry which occur together at 10 different locations in the site (Table VIII).

It appears then that there is a complex of 9 to 10 plants, the remains of which tend to be found together in the upper levels of the site (at a depth of about 1 foot, plus or minus 0.3 of a foot) and apparently represent occupation F. There are 9 instances of the occurrence of this complex from 8 squares distributed throughout the area from which plant remains were recovered. Hazelnut occurs in 2 of these, and corn and birchbark each occur with the complex in 3 cases. Three members of the complex occur together in 1 case, 4 in 2 cases, 5 in 2 cases, 6 in 2 cases, and 7 and 8 each in 1 case. Fire cherry occurs with the complex

TABLE VIII

ASSOCIATIONS OF PLANT REMAINS AT THE JUNTUNEN SITE

	Total Occurrences	Birchbark	Corn	Hazelnut	Blueberry	Pepperroot	Elderberry	Fire Cherry	Blackberry	*Galium*	Chenopod	Beechnut	Sumac Berry	Acorn	Bearberry	Grape	Cherry	Tuber?	Plum	Large Seed?	Rumex
Birchbark	27	10	8	10	1	1	3	4	4	3	2	2	1	2	2	. .	. .	3	2	1	. .
Corn	16	8	4	5	1	. .	2	4	3	2	1	1	2	2	1	. .	1	1	2	1	. .
Hazelnut	25	10	5	11	1	. .	1	3	2	2	2	1	1	1	2	. .	. .	2	2	1	1
Blueberry	1	1	1	1	. .	. .	. .	. .	. .	. .	. .	. .	. .	. .	. .	. .	. .	1	. .	. .	. .
Pepperroot	3	1	. .	. .	. .	1	2	2	2	. .	. .	1	1	. .	1	. .	1	. .	. .	. .	. .
Elderberry	8	3	2	1	. .	2	. .	8	7	6	5	3	3	3	2	1	1	. .	. .	. .	. .
Fire cherry	14	4	4	3	. .	2	8	1	10	7	5	3	3	4	4	1	. .	. .	. .	. .	. .
Blackberry	10	4	3	2	. .	2	7	10	. .	6	4	3	3	3	4	1	. .	. .	. .	. .	. .
Galium	7	3	2	2	. .	. .	6	7	6	. .	4	2	1	3	3	1	. .	. .	. .	. .	. .
Chenopod	6	2	1	2	. .	. .	5	5	4	4	. .	1	1	2	1	1	. .	. .	. .	. .	1
Beechnut	3	2	1	1	. .	1	3	3	3	2	1	. .	2	1	2	. .	. .	. .	. .	. .	. .
Sumac berry	4	1	2	1	. .	1	3	3	3	1	1	2	. .	1	1	. .	. .	. .	. .	. .	. .
Acorn	4	2	2	1	. .	. .	3	4	3	3	2	1	1	. .	1	1	. .	. .	. .	. .	. .
Bearberry	4	2	1	2	. .	1	2	4	4	3	1	2	1	1	. .	. .	. .	. .	. .	. .	. .
Grape	1	. .	. .	. .	. .	. .	1	1	1	1	1	. .	. .	1	. .	. .	. .	. .	. .	. .	. .
Cherry	3	. .	1	. .	. .	1	1	. .	. .	. .	. .	. .	. .	. .	. .	. .	. .	. .	. .	. .	. .
Tuber?	7	3	1	2	1	. .	. .	. .	. .	. .	. .	. .	. .	. .	. .	. .	. .	4	. .	. .	. .
Plum	2	2	2	2	2	. .	. .	. .	. .	. .	. .	. .	. .	. .	. .	. .	. .	. .	. .	. .	. .
Large seed?	7	1	1	1	. .	. .	. .	. .	. .	. .	. .	. .	. .	. .	. .	. .	. .	. .	. .	4	. .
Rumex	1	. .	. .	1	. .	. .	. .	. .	. .	. .	1	. .	. .	. .	. .	. .	. .	. .	. .	. .	. .

9 times, blackberry and elderberry each 8 times, *Galium* 7 times, bearberry and chenopod 4 times each, and beechnut, acorn, and sumac 3 times each. These are the total number of occurrences for elderberry, *Galium*, bearberry, and beechnut. The single grape seed occurrence is with 6 members of the complex, and the pepperroot tuber also occurs once with 6 members.

It is not the contention that in every case where the individual members of the complex occur occupation F is represented; but in no case where at least three or four of them occur together is there definite evidence of antiquity greater than occupation F. Another indication of this is that all of the plant remains known to be earlier are thoroughly charred with the exception of some of the birchbark. At least some of the fire cherry pits, chenopod seeds, blackberry seeds, elderberry seeds, and sumac seeds in most of the collections are not charred at all.

With only four exceptions, the known plant foods represented are available for collection in August. Pepperroot tubers may be collected in April to May, elderberries *(Sambucus pubens)* ripen from June to July, beechnuts ripen from October to November, and chenopod seeds ripen from early autumn to early winter. The indication is that most of the plant remains were deposited during the period of July to September and perhaps from May to October or a little longer. *Galium* seeds are mature during approximately the period of May to October, whereas *Rumex* seeds mature from June to September (Fernald, 1950). It may be that the pepperroot tubers were collected later or that they were stored and used in late summer, but there is nothing to indicate that the Indians were not present on the site as early as May. In fact, evidence from analysis of the prodigious quantities of fish bone present indicate that they were (Charles B. Cleland, personal communication, 1962). Beechnuts were stored for winter by the Menomini (Smith, 1923: 66), but presumably not for summer.

The Potawatomi apparently collected most of their beechnuts from deer mouse stores, but this collecting takes place in winter (Smith, 1933: 100). The Chippewa similarly made use of chipmunk stores of these nuts (Gilmore, 1933: 128). Thus, it would appear that the nuts were collected and the shells deposited as refuse as late as October and intermingled with debris that had been discarded earlier in the year. The animal bone analysis again would agree since it indicates an occupation from late spring until November.

An occupation lasting at least from May to October is supported also by climatic data from the *1941 Yearbook of Agriculture*. The average date of the latest spring frost is about May

18 while the average date of the first autumn frost is about October 6. This gives the Juntunen site an average frostless season of at least 140 days. Other available climatic data include the following approximations: average January temperature, 19 F; average July temperature, 66 F; average annual precipitation, 29 inches; average warm season precipitation (April through September), 17 inches.

Scarcity of mammal bones and projectile points at the site indicates that it was not occupied during winter (Charles B. Cleland, personal communication, 1962). This seems a reasonable supposition also because of the open location of the site on a northerly shore, exposing it to all the rigors of winter.

E. R. Biel (1961: 332-3) points out that:

> Foresters have always been aware of the great climatic differences between the interior and the surroundings of woodland . . . . Canopy interference and windbreaking effects of trees create special forest climates whose intensities vary with differences in canopy density (coniferous *versus* deciduous forests, seasonal variations in foliage, multi-layered tropical forests *versus* uniform-layered stands in forests under long cultivation, etc.) . . . the interference with respect to incoming and outbound radiation varies tremendously, and temperature conditions, evaporation rates, vapor, and dust contents are controlled by screening and stagnation.

Thus the forest would be a safer and more comfortable place to live during winter because it is better protected from the severities of this season and firewood is more easily available. It may be that the summer village population broke up into smaller family groups which lived here and there in favorable locations on the island or possibly even on the mainland. The island might be a more expedient place to overwinter because of the ameliorating influence of the surrounding lake waters and because game could be driven in any direction into the water or onto the open ice for easier slaughter.

There may have been one or more additional moves before the late spring reoccupation of the village site (when food was again more abundant), specifically perhaps to maple groves where the trees would be tapped for the sweet sap. Since early spring is likely to be the time of the most severe food shortage, the availability of maple sap at this time would be extremely important. Resettlement of the village would then take place in time for planting corn and resuming extensive fishing operations.

The Juntunen site apparently was well cleared at the time of occupation F. This is indicated by the soil and stratigraphy studies made by Henry Wright and also by the presence of weed seeds at that level. The weeds represented are two species of

Chenopodium, Galium, Rumex, and perhaps some of the unidentified seeds. They may have grown upon the site itself or perhaps were present mostly around the periphery. The *Galium* seeds are especially significant at this point because they were so prolific in the strata of occupation F. The *Galium* seeds are about 2.5 to 3.0 millimeters in diameter and hollow on one side. In the fresh state the fruits of *G. triflorum* Michx., the most likely of six candidates, are densely covered with bristles and thus would stick to clothing, etc. and be brought into the village area from its peripheries or cornfields if it did not actually grow on the site. *Galium triflorum* (fragrant bedstraw) has been used for bedstraw as the name indicates, and the Menomini used it to brew a medicinal tea (Smith, 1923: 51). *Galium aparine* L., *G. trifidum* L., and *G. tinctorium* L. were used medicinally by the Ojibwa (Smith, 1932: 386-87); but it is unlikely that the archaeological seeds are of these species.

Perhaps the most interesting plant remains found, at least from a strictly botanical point of view, are the *Chenopodium* seeds. These are referable to two species, *C. album* L. and *C. hybridum* L. var. *gigantospermum* (Aellen) Rouleau, according to my identifications. Dr. E. H. Moss, Professor Emeritus of Botany of the University of Alberta, examined the same seeds. Dr. Moss (personal communication, 1962) reported that the large collection of smaller seeds "does appear to be *Chenopodium album* L. (in the broad sense, and including *C. lanceolatum* Muhl.)," and he suggested that the larger seeds are *C. hybridum* L. var. *gigantospermum* (Aellen) Rouleau. The larger chenopod seeds occurred in only one collection from the Juntunen site at a depth of about 8 inches. Its appearance there may be only as a natural deposit from a weed, but it occurs with fire cherry and blackberry seeds.

The most significant aspect of the occurrence of the seeds of *C. album* is that this species was naturalized from Europe according to Fernald (1950: 595) and other authors. Deam (1940: 423), however, comments as follows: "Most American plants identified as *C. album* actually belong to *C. Berlandieri* ssp. *Zschachei* (Murr) Zobel. Some specifically American races, however, are found in addition to the truly European races introduced all over the world. The plants are found in sandy soils along roadsides." Fernald (1950: 595) gives the range of *C. berlandieri* Moq. as entirely south and west of the Upper Great Lakes region. Alexander Johnston (1962: 129) cites a personal communication from E. H. Moss that states as follows in reference to *C. album:* "Although the species is commonly thought to be introduced, it is probably also native to Alberta, Canada."

"Recently a cache of seeds, identified as *C. album*, was found at a depth of 1.5 m. near the Ross site, an archaeological 'dig' along the Oldman River northeast of Coaldale, Alberta" (Johnston, *ibid.).* The cache, containing four to five quarts of seeds, was from a relatively late occupation; but there were no European trade goods associated. Blankinship (1905) states that seeds of *C. album* were used for food by the Indians of Montana "which may account in part for its wide distribution and abundance in the state as is probably the case with the sunflower *Helianthus annus* L. and the horseweed *Iva xanthifolia* Nutt. . . ." In the Upper Great Lakes region the Potawatomi and Ojibwa used the young tops of *C. album* for food (Smith, 1933: 47; Reagan, 1928: 24), and the Ojibwa used chenopod seeds to make flour (Stowe, 1940: 12).

We have found no records of the use of any other small seeds for food in the Great Lakes regions. It is interesting in this connection that no grass seeds were found in the Juntunen site, not even wild rice which holds up well under burning and washing. Wild rice, however, has never been definitely identified from a prehistoric archaeological context, to my knowledge.

The Juntunen corn is the northernmost archaeological occurrence, with the exception of corn from relatively late sites in North Dakota. No parts of the plant were found other than kernels, but these were found throughout the area of the site which produced remains of plant food and at lower levels than any other plant except cherry, hazelnut, and birchbark. The corn is of the race called Eastern Complex but is a rather generalized (and perhaps early) form. It may be the earliest Michigan corn with the exception of that found at the Younge site in southeastern Michigan which has tendencies toward the apparently earlier "Old Village" type of Middle Mississippian corn (Plate I).

Though most of northern Lower Michigan is within the zone of less than 140 days of frostless season, there is a narrow band with a longer frost-free period that extends along the northwest coast of the Lower Peninsula to the eastern tip of Upper Michigan. This is probably the route by which the Juntunen inhabitants received corn; and it seems likely that older corn might be found somewhere between Mackinac City and Grand Rapids, perhaps in the Traverse Bay area. However, corn may have been carried to Bois Blanc Island from the Green Bay area of Wisconsin.

In summary, the remains of 15 plant foods were identified from materials excavated at the Juntunen site, most of which came from the latest level of occupation. They include 9 fruits and berries, 3 nuts, chenopod, a tuber, and corn. In addition to this, birchbark was found at many locations throughout the site; and 2 kinds of weed seeds were identified. Common plum, common hazelnut, paw paw, chestnut, butternut, walnut, and hickory nut have all been found archaeologically in the Midwest; but their ranges do not reach northward as far as Bois Blanc Island.

Later chapters give a fuller accounting of the significance of the native plant food remains, the corn typology, and the location of the Juntunen site relative to climatic factors and agriculture prehistory in the Midwest.

III

NATIVE PLANT UTILIZATION

Indian utilization of at least 373 plants native to the Upper Great Lakes region is indicated in the ethnobotanical literature. The products of about 130 plants were used for food, 18 for beverage and flavoring, 68 for medicinal teas, 207 for other medicines, 31 for charms and ceremonial purposes, 27 for smoking, 25 for dyeing, and 52 for various utilitarian purposes. Those used for food and utility were by far the most important to aboriginal subsistence. These include 167 plants, since 15 plants were used both for food and for utility. Plants in the nonfood categories are listed with annotations in the appendices.

In several cases it is not known how many species within a particular genus were actually utilized, that is, thorn apple, blackberry, grape, oak, chenopod. Probably all available members of these genera were used to some extent. Thus the actual number of species used would be somewhat higher than the figures given here. Also there are probably a number of plants whose uses have been forgotten or simply have not been reported. Thus 400 to 500 native plant species perhaps were utilized aboriginally in the Upper Great Lakes region alone.

It is my estimation that the total number of native vascular plant species in the Upper Great Lakes region approximates 2000. This estimate was made by counting the species listed in *Gray's Manual of Botany* (Fernald, 1950) as occurring in this region. The counting proceeded one page at a time at regular intervals throughout the book until more than 8 per cent of all the listed species were counted. A check of the results reveals that 159 of the 445 plants counted are native to the Upper Great Lakes region. Multiplying this ratio by 5523, the total number of plant species listed, indicates that 1973 flowering plants, ferns, and fern allies are native to the region. Thus at least 20 per cent of the available vascular plants apparently were used by the Indians aboriginally, and of these more than one-third were used for food.

Altogether, there are about 560 uses recorded for the 373 plant species utilized. Thus there is much overlapping between the categories of plant utilization. The greater the number of species used for one purpose, the smaller is the percentage of overlap. The food plants and medicinal plants account for 58 per cent of the overlaps in use and 72 per cent of the total uses. Table IX shows the overlap for each category of use and the number of plant families and genera represented in each category. Table X lists the families with the greatest numbers of collective uses.

Two plant species, bearberry and staghorn sumac, are reported to have uses in five of the categories. Six plants, including hemlock, sweet fern, hazelnut, hawthorn, smooth sumac, and milkweed, have four uses each. There are three uses each for 28 plants, two for 102 plants, and one for 235 plants.

The 4425 native plants listed in *Gray's Manual of Botany* (Fernald, 1950) may be divided as follows: 120 pteridophytes (rush-like, fern-like, moss-like or quill-leaved plants without true flowers), 2.7 per cent; 26 gymnosperms (conifers), 0.6 per cent; 1250 monocotyledons (grasses, sedges, rushes, lilies, orchids, etc.), 28.2 per cent; and 3029 dicotyledons (all other higher plants), 68.5 per cent. Presumably the percentage distribution is essentially the same among the approximately 2000 native species of the Upper Great Lakes region. The distribution of "economic plants" among these same groups is as follows: 14 pteridophytes, 3.6 per cent; 12 gymnosperms (of 13 in the Upper Great Lakes region), 3.3 per cent. Thus the percentages of gymnosperms and dicotyledons used is somewhat higher than might be expected and the percentage of monocotyledons is somewhat lower. Only seven of the grasses and sedges are recorded as having been utilized, but these two groups comprise more than two-thirds of the monocotyledons listed in *Gray's Manual of Botany*. However, wild rice *(Zizania aquatica*, a grass) and the great bullrush *(Scirpus validus*, a sedge) apparently were rather important to the aboriginal economy. Monocotyledons were not used at all for dyeing, smoking, or beverage, but they constitute more than a third of the charm plants. Pteridophytes were used only for food and medicine. All charm plants are angiosperms, and only dicotyledons were used in smoking.

Native Food Plants

The list of food plants that follows is composed of the ones used by Indians who lived in the Upper Great Lakes region when they were first contacted by Europeans and who supposedly lived there in prehistoric times.

Since only a few of these plants are represented by archaeological identifications from the Upper Great Lakes region (at least

TABLE IX

PLANT TAXA ACCORDING TO UTILIZATION AND USE OVERLAP IN THE UPPER GREAT LAKES REGION

Use	Species	Genera	Families	Collective Number of Other Uses	Per Cent Overlap
Beverage and flavoring	18	16	11	34	189
Dye	25	21	15	41	164
Smoking	27	18	12	36	133
Charm	31	30	19	36	116
Utility	52	38	25	58	112
Food	130	80	48	114	88
Medicine	275	183	71	117	62

TABLE X

PLANT FAMILIES MOST UTILIZED IN THE UPPER GREAT LAKES REGION

	Minimum No. of Species Used	Per Cent of Species in Fernald (1950)	Collective Number of Uses	Predominant Uses
Compositae . . .	43	7.6	52	36 medicine 8 smoking
Rosaceae	36	7.4	55	25 medicine 23 food
Liliaceae	16	18.8	24	12 medicine 9 food
Ericaceae	15	21.4	26	8 medicine 11 food
Ranunculaceae	14	15.9	20	13 medicine 4 dye
Pinaceae.	11	44.0	25	11 medicine 10 utility

20 of perhaps 40 possibilities from about a dozen sites), it cannot definitely be stated that the list is representative of the wild food plants utilized more than a few hundred years ago. The climate probably has remained much the same for the past 4000 years, however, except for relatively minor fluctuations; and the general vegetational patterns probably have not changed much during this period. This is supported by the fact that the tree genera which were the sources of the remains of seeds and charred wood from the 4000-year-old Feeheley site near Saginaw (discussed above) are still abundant in the same area.

It is unlikely that an appreciable number of plant species have died out in the area during the four millenia preceding the European invasion of America or that many have entered from the north, but some food plants probably entered from the south and perhaps also from the east or west during the period. Nevertheless, it seems likely that most of the food plants available to the protohistoric inhabitants of the Saginaw area, for instance, were also available to the people who buried their dead at the Feeheley site. We can be reasonably certain, however, of utilization only for those plant species whose identified remains have been recovered archaeologically and which evidently were not intrusive.

The list of aboriginal food plants may be somewhat shorter than it should be because of abandonment of some plant resources during the historic period. Also, the historic importance of some plants may not accurately reflect their aboriginal importance which may have increased or decreased as a result of social and technological upheaval. Trade goods have long since replaced much of the wild plant produce. The activities of men involved in the fur trade may have increased the dependence upon plant-food gathering in order to offset the loss of foods resulting from neglect of hunting, fishing, and agriculture.

Most of the ethnobotanical data available for Upper Great Lakes Indians have been obtained from the Ojibwa, Menomini, Sauk-Fox, and Potawatomi, all Algonkian. Very little information of this type is available for the Winnebago, Kickapoo, Mascouten, Ottawa, Huron, Tobacco, Neutral, or Miami. Iroquois data are included whenever possible to substitute in part for the absence of data from the southern and eastern parts of the region.

One plant food is listed for Ottawa, 3 for Algonkin, 4 each for Huron and Winnebago, 7 for Mascouten, 9 for Kickapoo, 29 for Sauk-Fox, 39 for Potawatomi, 45 for Menomini, 55 for Iroquois, and 89 for Ojibwa is due to the better preservation of their aboriginal subsistence base and to the greater amount of data available from four separate ethnobotanical studies (Densmore, 1928; Gilmore, 1933; Reagan, 1928; Smith, 1932). There is only one such comparable study each for Sauk-Fox, Menomini, and Potawatomi (Smith, 1928, 1923, and 1933) and none for Ottawa, Algonkin, Kickapoo, Winnebago, Huron, Tobacco, and Neutral.

In actuality, the number of plant foods used by the Ojibwa probably was among the lowest in the Great Lakes region. The Iroquoian tribes probably used a somewhat greater number of wild food plants, even though they were agricultural; because more of these plants were available to them. For instance, many nut- and fruit-bearing trees grow in the former Iroquoian territories that are essentially absent in the Ojibwa territory; whereas very few food plants available to the Ojibwa were unavailable to the Iroquoians. Many food plants used by the Iroquois are not included in the accompanying list because they do not occur, or occur only sparingly, within the Upper Great Lakes region.

The Latin name of each food plant is followed by the common name or names. Following this in some cases is the designation *N* or *S* indicating a generally northern or southern distribution of the plant. The plant product utilized is followed by the season at which it is available and, in some cases, the frequency of good crop years. This information was compiled mostly from the following sources: Billington (1943), Fernald (1950), Fernald and Kinsey (1943), Forest Service (1948), Forestry Branch (1956), Medsger (1939), Otis (1931), Peterson (1923), and the U.S. Department of Agriculture (1961).

Geographical range and habitat preference data are from Billington (1943), Deam (1940), Fernald (1950), Fernald and Kinsey (1943), Fink and Hedrick (1935), Forestry Branch (1956), Little (1949), Medsger (1939), Nearing (1947), Otis (1931), and files prepared by Dr. Edward G. Voss at the University of Michigan Herbarium. Only north--south range is presented except in a few cases. Following this in many cases is the designation *IR* or *K* or both, indicating the presence of the species on Isle Royale (Brown, n.d.) or in Kalamazoo County in southwest Michigan (Hanes, 1947).

Archaeological occurrences of remains of the species are added when there are records of them. Those from Ohio are from Goslin (1952, 1957, and personal communication, 1963) unless otherwise credited. Those from Michigan not otherwise credited were identified by the writer.

The penultimate item, where applicable, is the record of my observation of occurrence of the plant growing at Late Woodland archaeological sites. The sites included are the Mikado, Rifle

River, Linseman–Walters, and Missaukee Earthwork enclosures in Alcona, Ogemaw, and Missaukee counties, two sites near the Manistee River in Wexford County, and the Juntunen site on Bois Blanc Island, all in Michigan.

The final item in each entry, if it occurs, is the nonfood uses of the plant.

Food Plants in Order of Availability by Season

Early Spring

1. *Acer saccharum* Marsh. Sugar maple (including A. *nigrum* Michx.).

Sap runs about March or April depending on weather. Ojibwa went from winter camps to sugar camps in early April for a month (Smith, 1932: 394). Used by Ojibwa, Menomini, Potawatomi, Sauk-Fox, Winnebago, and Iroquois (Smith, 1932: 394; 1923: 61; 1933: 93; Henshaw, 1890). Grows from Alabama to northern Lake Superior in moist, rich soil in valleys and uplands and on moist rocky slopes. IR, K. Found growing at a Rifle River enclosure. Also used to make arrows, paddles for stirring food, bowls, etc.

2. *Acer saccharinum* L. Silver maple. S.

Sap runs about March or April. Used by Ojibwa and Menomini (Gilmore, 1933: 136; Reagan, 1928: 234; Hoffman, 1896: 237). Grows from the Gulf of Mexico to Lake Superior and southern Ontario in low, rich bottom land with occasional flooding, but not in swamps. K. Found growing at a Rifle River enclosure. Also used for medicine.

3. *Acer negundo* L. Box elder. S.

Sap runs about March or April. Used by Ojibwa (Smith, 1932: 394). Grows from the Gulf of Mexico to southern Ontario, Saginaw Bay, and southeastern Minnesota along stream banks and swamp borders in deep, moist soil. K. Also used for medicine.

4. *Betula lutea* Michx. Yellow birch. N.

Sap available anytime. Used by Ojibwa and Potawatomi (Smith, 1932: 397; 1933; 44). Grows from northern Ohio and northern Illinois to central Ontario (larger and more frequent in the north) in rich, moist upland woods. IR, K. Found growing at a Rifle River enclosure. Also used as a medicine seasoner and for wigwam poles.

5. *Gaultheria procumbens* L. Teaberry, checkerberry, wintergreen.

Berries ripen in the fall but stay on all winter and are larger in early spring. Used by Ojibwa and Iroquois (Smith, 1932: 400;

Waugh, 1916: 128). Grows from Kentucky to central Ontario in thin sandy woods and clearings (common throughout Michigan). IR, K. Found at a Rifle River enclosure. Also used for beverage, flavoring, and medicine (contains methyl salicylate).

6. *Allium cernuum* Roth. Wild onion. S.

Bulbs available in late autumn and early spring. Used by Ojibwa (spring food) and Menomini (Smith, 1932: 406; Skinner, 1921: 64). Grows from Tennessee to New York, central Lower Michigan, and British Columbia on ledges, gravels, and rocky or wooded slopes and crests. K.

7. *Allium tricoccum* Ait. Wild leek, ramp.

Bulbs available in late autumn and early spring. Used by Ojibwa and Menomini (spring food and dried for future use), (Smith, 1932: 406; 1923: 69; 1933: 104; Waugh, 1916: 118). Grows from Kentucky to southern Ontario, Lake Superior, and Minnesota in rich woods and bottomlands. K. Found growing at a Wexford County site. Also used for medicine.

8. *Scirpus validus* Vahl. Great bulrush.

Tubers available in autumn and early spring; stem-base available in May to June. Used by Ojibwa and Iroquois (Densmore, 1928: 293, 307; Parker, 1910: 105). Grows from Mexico to Hudson Bay in brackish or fresh shallow water and marshes. IR, K. Found archaeologically in textiles in Ohio at Ash Cave, Canter's Caves, and South Park Village (UMMAEL Repts. 15, 33, and 71).[1] Stems used to make mats and baskets.

Spring

9. *Nelumbo lutea* (Willd.) Pers. *Wonkapin*, water chinquapin, American lotus. S.

Tubers available in spring and autumn; nuts available in late summer and autumn. Used by Ojibwa, Potawatomi, and Sauk-Fox; dried and stored for winter (Smith, 1932: 407; 1933: 105; 1928: 262). Grows from the Gulf of Mexico north locally to southern Ontario, Minnesota, and the Great Lakes (where it is most abundant around the west end of Lake Erie, especially around the mouth of the Maumee River, and in Lake Winnebago) in ponds, quiet streams, and estuaries. K. Perhaps introduced generally or locally by the Indians.

[1] University of Michigan Museum of Anthropology, Ethnobotanical Laboratory Report. These reports are on file in the Laboratory.

10. *Nuphar advena* (Ait.) Ait. Cow lily, yellow pond lily. S.
Tubers available in autumn to early spring. Used by Sauk-Fox, Menomini, Mascouten, and Iroquois (Skinner, 1925: 138; Smith, 1923: 69; Skinner, 1926: 286; Parker, 1910: 105). Grows from Mexico to southern Michigan and Wisconsin in pond margins and swamps. K. Also used for medicine.

11. *Claytonia virginica* L. Spring beauty. S.
Tubers available in spring (April to June). Used by Iroquois (Waugh, 1916: 120). Grows from Georgia to Minnesota, central Lower Michigan, and southern Ontario (*C. caroliniana*, with a tuber half as thick, grows from central Lower Michigan northward) in rich woods, thickets, and clearings. K.

12. *Osmunda cinnamomea* L. Cinnamon fern, buckthorn, fiddle heads.
Fiddle heads (immature fronds) available in spring when beginning to sprout. Used by Menomini (Smith, 1923: 70). Grows from the Gulf of Mexico to central Ontario and throughout the eastern United States and southeastern Canada in swamps, low woods, and thickets. IR, K.

13. *Pteridium aquilinum* (L.) Kuhn. Bracken fern.
Young sprouts available in spring (May to June). Used by Ojibwa in soup (Smith, 1932: 408). Grows from Mexico to the Great Lakes region in dry, open woods and recently burned clearings. IR, K. Found growing at the Wexford County and Hubbard Lake sites and at the Rifle River, Linseman--Walters, and Missaukee enclosures. Also used for medicine.

14. *Ulmus rubra* Muhl. Slippery elm, red elm. S.
Cambium easier to collect in spring. Used by Iroquois (Hunter, 1823*a*). Grows from Alabama to southern Ontario, northeastern Lake Superior, northwestern Michigan (frequent throughout the state), and northern Wisconsin on streambanks, bottomlands, rich, moist hillsides, and rocky ridges and slopes. K. Bark used for fiber and for making boxes, baskets, and wigwam sides; also used for medicine.

15. *Populus grandidentata* Michx. Large-toothed aspen.
Cambium probably collected in spring. Used by Ojibwa (Smith, 1932: 387). Grows from Tennessee to southern Ontario and the northern border to the Great Lakes region (common in northern Lower Michigan but rare in Upper Michigan) in rich, moist, sandy soil, swamp borders, riverbanks, and hillsides. IR, K. Also used for medicine.

16. *Fraxinus pennsylvanica* Marsh. Red ash.

Cambium probably collected in spring. Used by Ojibwa (Smith, 1932: 407). Grows from the Gulf of Mexico to northern Lake Superior (not common in Michigan, mostly in southern Lower Michigan) in wet or moist, rich loam, river banks, and swampy lowlands. K. Found growing at the Mikado enclosure. Wood used to make bows, arrows, snowshoes, sleds, spoons, splint baskets, etc.; also used for medicine.

17. *Tilia americana* L. Basswood, *bois blanc.*

Young twigs, buds, and sap available in spring. Used by Ojibwa and Iroquois (Densmore, 1928: 307; Gilmore, 1933: 137; Parker, 1910: 104). Grows from Tennessee to central Ontario (common in Lower Michigan; frequent in Upper Michigan) in rich, well-drained, loamy soils. K. Found archaeologically in textiles from Seip and Harness mounds and Ash and Canter's Caves in Ohio (Whitford, 1941: 11) and as carbonized bark in the Juntunen site in Michigan. Found growing at the Mikado, Rifle River, and Missaukee enclosures. Inner bark bast used for cordage (very important to subsistence); trunks used to make dugout canoes.

18. *Pinus strobus* L. White pine. N.

Young staminate catkins available in spring, inner bark in May and early June. Used by Ojibwa and Iroquois (Smith, 1932: 407; Waugh, 1916: 119). Grows from southern Michigan and northern Illinois through Wisconsin and southern Ontario to James Bay in light fertile loam and sandy soils of granitic origin. IR, K. Found growing at the Rifle River and Missaukee enclosures. Pitch used for caulking and waterproofing; wood used to make dishes; also used for medicine.

19. *Dentaria laciniata* Muhl. Pepperroot, toothwort, crowfoot (including D. *maxima* Nutt.).

Tubers available in April to May. Used by Ojibwa, Menomini, and Iroquois (Smith, 1932: 399; 1923: 65; Waugh, 1916: 120). Grows from the Gulf of Mexico to central Ontario in rich woods, wooded bottoms, and calcareous rocky banks preferring moist soil and deep leafmold, frequently occurring in large colonies. K. Carbonized tubers found at two locations in the Juntunen site in Michigan.

20. *Dentaria diphylla* Michx. Pepperroot, toothwort. S.

Tubers available in April to May. Used by Iroquois (Waugh, 1916: 120). Grows from southern Ohio to northeastern Wisconsin and southeastern Upper Michigan in damp, rich woods, moist and sandy soil, and low alluvial flats, frequently occurring in large colonies. K.

21. *Apios americana* Medic. Groundnut, rosary root, *pin*. S.
Tubers available year round? Gathered by Ojibwa in spring (Densmore, 1929: 123); dried for winter by Menomini (Smith, 1923: 68); sometimes planted by Iroquois (Waugh, 1916: 120); used by Potawatomi, Mascouten, Sauk-Fox, and Kickapoo (Smith, 1933: 103; Skinner, 1926: 286; Smith, 1928: 259; Hunter, 1823*b*: 265; Beardsley, 1940: 509). An important food for all tribes within its range; great quantities once grew along the Pinconning ("place of the *pin*") River in Bay County, Michigan, where the Indians returned every year (Gilmore, 1933: 133). Grows from the Gulf of Mexico to southern Ontario and Lake Superior (mostly in the southern half of Lower Michigan) in rich, damp thickets and boggy land in abundance. "It was formerly common in damp or marshy soil throughout the woodlands and prairies of eastern and central temperate North America" (Beardsley, 1940: 507).

22. *Arisaema triphyllum* (L.) Schott. Indian turnip, jack-in-the-pulpit.
Tubers available in spring and late autumn. Used by Ojibwa, Potawatomi, and Iroquois (Smith, 1932: 95; 1933: 95; Parker, 1910: 105). Grows from the Gulf of Mexico to James Bay in rich, low woods, chiefly near streams. IR, K. Also used for medicine.

23. *Aralia racemosa* L. Spikenard.
Young tips available in spring. Used by Potawatomi in soup (Smith, 1933: 96). Roots used by Menomini (Smith, 1923: 62). Grows from Alabama to central Ontario in rich woods and thickets. K. Also used for medicine ("one old [Ojibwa] man cultivates a patch of this plant" [Reagan, 1928: 237]).

Late Spring

24. *Caltha palustris* L. Marsh marigold, cowslip. N.
Available for greens in late spring. Used by Ojibwa, Menomini, and Iroquois (Smith, 1932: 408; 1923: 70; Waugh, 1916: 117). Grows from southern Ohio to northern Canada in swamps, wet woods, and wet meadows. IR, K. Also used for medicine.

25. *Asclepias tuberosa* L. Butterfly weed. S.
Shoots available in late spring and early summer. Used by "some Indians of Canada" (Palmer, 1871: 405). Grows from the Gulf of Mexico to northern Lower Michigan, northern Wisconsin, and southern Ontario in dry, open soil. K. Archaeological fibers found in an Ohio mound textile (Whitford, 1941: 10). Also used for medicine.

26. *Asclepias syriaca* L. Common milkweed.

Shoots available in late spring and early summer, buds and flowers in early summer, and young pods in summer and early autumn. Shoots used by Ojibwa, buds and flowers by Ojibwa, who dry them for winter food, and by Menomini, Potawatomi, Winnebago, Sauk-Fox, and Iroquois (Smith, 1932: 397; 1923: 62; 1933: 96; Radin, 1923: 118; Smith, 1928: 256; Waugh, 1916: 117). Grows from Alabama to James Bay in thickets, dry fields, and roadsides. K. Archaeological fiber found in Ash and Canter's Caves in Ohio (Whitford, 1941: 10). Found growing at a Wexford County site, at the Rifle River and Missaukee enclosures, and at the Juntunen site. Also used for fiber, hunting charm, and medicine.

27. *Heracleum maximum* Bartr. Cow parsnip, masterwort. N.

Available for greens in late spring and early summer. Used by Ojibwa (Reagan, 1928: 237). Fox use the root for food (Smith, 1928: 265). Grows from southern Ohio to central Hudson Bay in rich or low ground and alluvial thickets. IR, K. Also used for hunting and fishing charm and medicine.

28. *Hydrophyllum virginianum* L. John's cabbage, Virginia waterleaf. S.

Available for greens in late spring. Used by Menomini and Iroquois (Smith, 1923: 68; Waugh, 1916: 117). Grows from Tennessee to southern Ontario, southern Upper Michigan, and northern Wisconsin in rich woods and damp clearings. K. Also used for medicine.

29. *Oxalis montana* Raf. Common wood sorrel, wood shamrock. N.

Available for greens in late spring and summer. Used by Potawatomi and Iroquois (Smith, 1933: 68; Parker, 1910: 93). Grows from southern Michigan and southern Wisconsin to James Bay in damp woods.

30. *Allium canadense* L. Wild onion. S.

Bulbs available in late spring. Used by Menomini, Potawatomi, Sauk-Fox, and Iroquois (Smith, 1923: 69; 1933: 104; 1928: 262; Waugh, 1916: 118). Used in soup and considered a valuable wild food. Grows from the Gulf of Mexico to southern Ontario, central Lower Michigan and Minnesota in low woods, thickets, and meadows, K.

31. *Populus tremuloides* Michx. Quaking aspen. N.

Sap available in early summer. Used by Ojibwa (Densmore, 1928: 291, 320). Grows from southern Ohio to Hudson Bay (common throughout Michigan) in moist, sandy soil, gravelly hillsides, and burned-over forest. IR, K. Bark and bast fibers found archaeologically in Ash and Canter's Caves in Ohio (UMMAEL Rept. 15). Found growing at the Wexford County sites and at the Mikado, Rifle River, and Missaukee enclosures. Also used for medicine.

32. *Chenopodium* spp. *(C. album?)* Goosefoot, lambsquarters, pigweed, chenopod.

Available for greens in early summer and seeds in autumn to early winter. Greens used by Ojibwa, Potawatomi and Iroquois; seeds used by Ojibwa (Stowe, 1940: 12; Smith, 1933: 47; Waugh, 1916: 117). Grows throughout the Great Lakes region and beyond in cultivated and waste ground. IR, K.

33. *Aster macrophyllus* L. Large-leaved aster, starwort. N.

Available for greens in early summer. Used by Ojibwa (Smith, 1932: 398). Grows from southern Ohio to central Ontario in dry to moist, open woods, thickets, and clearings. IR, K. Found growing at a Rifle River enclosure. Also smoked to attract deer; used for medicine.

34. *Fragaria virginiana* Duchesne. Wild strawberry.

Fruit available in early summer. Used by Ojibwa (important food in season; preserved for winter), Menomini, Sauk-Fox, Kickapoo, and Iroquois (Gilmore, 1933: 132; Smith, 1932: 409; 1923: 71; 1928: 263; Hunter, 1823*b*: 265; Waugh, 1916: 127). Grows from the Gulf of Mexico to Hudson Bay abundantly in fields, dry meadows, open slopes, and borders of woods. Low in food value (Hinsdale, 1932: 9), but the earliest abundant fruit. IR, K. Archaeological achenes found with sunflower and marshelder seeds in feces from Salts Cave, Kentucky. Found growing at a Rifle River enclosure. Also used for medicine.

35. *Fragaria vesca* L. var. *americana* Porter. Sow-teat strawberry. N.

Fruit available in early summer. Much used by Potawatomi (sometimes dried) and Ojibwa in season (Smith, 1933: 107; Gilmore, 1933: 132) and Iroquois (Waugh, 1916: 127). Grows from southern Ohio to James Bay in rocky woods and openings. K. Also used for medicine.

36. *Sambucus pubens* Michx. Red-berried elder, stinking elder. N.

Berries available in June to July. Used by Ojibwa (Reagan, 1928: 237) but inedible according to Fernald and Kinsey (1943: 351). Grows from southeastern Ohio and northern Illinois to

James Bay (common throughout Michigan) in rocky woods and dry shade. IR, K. Seeds occurred at seven locations in the Juntunen site in Michigan. Found growing at the Juntunen site. Also used for medicine.

37. *Asclepias incarnata* L. Swamp milkweed. S.
Flower buds available in early summer. Used by Menomini (Smith, 1923: 62). Grows from Florida to northern Upper Michigan in swamps, wet thickets, and shores. IR, K. Fiber found in archaeological textiles from Ohio; used for medicine.

Summer

38. *Nymphaea odorata* Ait. Pond lily, fragrant water lily.
Flower buds available in June to September. Used by Ojibwa (Smith, 1932: 407). Grows throughout the eastern United States and southeastern Canada to James Bay in ponds, dead waters, and bog pools. IR, K. Also used for medicine.

39. *Maianthemum canadense* Desf. Two-leaved Solomon's-seal, scurvy berries. N.
Berries available in summer through winter. Used by Potawatomi (Smith, 1933: 105). Grows from southern Ohio to Hudson Bay in woods and recent clearings. Found growing at the Rifle River and Missaukee enclosures. Also used for medicine.

40. *Amelanchier canadensis* (L.) Medic. Juneberry, service berry, shadbush. S.
Berries available in July to August. Used by Ojibwa, Menomini and Iroquois; dried for winter use; an important food (Gilmore, 1933: 132; Smith, 1923: 70; Waugh, 1916: 128). Grows from Georgia to northern Michigan and southern Ontario in moist woods and rocky uplands. Also used for medicine.

41. *Amelanchier laevis* Wieg. Juneberry, etc. N.
Berries available in July to August. Used by Ojibwa and Menomini; dried for winter use (Smith, 1932: 408, 1923: 70). Grows from southern Ohio to central Ontario in dry to moist thickets, borders of woods, margins of swamps, and clearings. K. Also used for medicine.

42. *Amelanchier sanguinea* (Pursh) D.C. Juneberry, etc. N.
Berries available in late July and August. Used by Potawatomi and other Wisconsin tribes; dried for winter use (Smith, 1933: 107). Grows from northern Ohio to James Bay in open woods, rocky slopes, river banks, etc.

43. *Amelanchier huronensis* Wieg. Juneberry, etc.
Berries available in July to August. Used by Menomini; dried for winter use (Smith, 1923: 70). Grows from the Manitoulin district to the Thunder Bay district south to the Bruce Peninsula, northern Michigan, Wisconsin, and Minnesota in open woods, cliffs, and shores. IR.
Some other Juneberries probably were used for food. *A. stolonifera* was found growing at the Juntunen site in Michigan.

44. *Gaylussacia baccata* (Wang.) K. Koch. Black huckleberry.
Berries available in July to August. Used by Ojibwa and Iroquois (Reagan, 1928: 238; Waugh, 1916: 128). This is among the richest of fruits (Hinsdale, 1932: 9). Grows from Alabama to central Ontario (frequent throughout Michigan) in rocky woodlands, swamps, and bogs. K.

45. *Ribes triste* Pall. Red currant. N.
Berries available in July to August. Used by Ojibwa and Iroquois; dried for winter use (Smith, 1932: 389; Waugh, 1916: 128). Grows from central Lower Michigan to central Hudson Bay (not in southern Wisconsin nor the southern tip of Ontario but well distributed in Michigan) in cold woods, swamps, and bogs. IR. Also used for medicine.

46. *Ribes hudsonianum* Richards. Currant. N.
Berries available in July to August. Used by Ojibwa (Reagan, 1928: 236). Grows from central Lower Michigan to central Hudson Bay and Alaska in swampy woods and rocky slopes.

47. *Ribes americanum* Mill. Wild black currant. N.
Berries available in July to September. Used by Ojibwa, Sauk-Fox, and Iroquois; dried for winter (Gilmore, 1933: 131, 410; Smith, 1928: 264; Waugh, 1916: 128). Grows from southern Ohio to central Ontario (common throughout Michigan) in rich thickets and slopes. K.

48. *Ribes* sp. Wild currant.
Berries available about July to August. Used by Menomini (Morse, 1822: 52). Found growing at a Rifle River enclosure.

49. *Ribes oxyacanthoides* L. Smooth gooseberry, northern gooseberry. N.
Berries available in July to August. Used by Ojibwa (Smith, 1932: 410). Grows from northern Michigan to Hudson Bay in woods and low grounds. IR. Also used for medicine.

50. *Ribes missouriense* Nutt. Missouri gooseberry. S.
Berries available in July to September. Used by Ojibwa (Gilmore, 1933: 131). Grows from Tennessee to northeastern Lower Michigan in open woods and thickets.

51. *Ribes cynosbati* L. Prickly gooseberry, dogberry.
Berries available in late July to September. Used by Ojibwa and Potawatomi; staple berry for Menomini, stored in winter (Gilmore, 1933: 131; Smith, 1933: 82; 1923: 71). Grows from Kentucky to central Ontario (common throughout Michigan) in open loamy or rocky woods, thickets, and hillsides. K. Also used for medicine.

52. *Ribes* sp. Gooseberry.
Berries available about July to September, Used by Huron and Iroquois (Raddisson, 1885: 143; Waugh, 1916: 127). Found growing at a Rifle River enclosure and at a Wexford County site.

53. *Rubus pubescens* Raf. Dwarf raspberry, plumboy, eyeberry. N.
Berries available in July to August. Used by Ojibwa and Iroquois (Gilmore, 1933: 133; Waugh, 1916: 127). Grows from southern Ohio to central Hudson Bay in damp slopes, rocky shores, low thickets, etc. K.

54. *Rubus odoratus* L. Purple-flowered raspberry, thimbleberry. S.
Berries available in July to September. Used by Ojibwa and Iroquois; dried for winter (Gilmore, 1933: 133; Waugh, 1916: 127). Grows from Tennessee to southern Ontario and northern Lower Michigan in thickets and borders of woods. *R. parviflorus* is closely related and grows from the Bruce Peninsula, northern Lower Michigan, and South Dakota to Alberta and Alaska.

55. *Rubus strigosus* Michx. Raspberry. N.
Berries available in July to August. Used by Ojibwa, Menomini, Potawatomi, Mascouten, and Iroquois; dried and stored for winter (Gilmore, 1933: 132; Smith, 1923: 71; 1933: 109; Skinner, 1926: 285; Waugh, 1916: 127). Grows from northern Ohio and northern Illinois to northern Canada in thickets, clearings, and borders of woods. IR, K. Found growing at a Rifle River enclosure. Also used for medicine.

56. *Rubus occidentalis* L. Black raspberry. S.
Berries available in July to August (good crop almost annually) Used by Ojibwa, Menomini, Sauk-Fox, and Iroquois; dried for winter (Gilmore, 1933: 133; Smith, 1923: 71, 1928: 264; Waugh,

1916: 127). Grows from the Gulf of Mexico to northern Lower Michigan, Minnesota, and southern Ontario (common throughout Michigan) in rich thickets, ravines, borders of woods, and waste ground. Also used for medicine.

57. *Rubus* spp. Dewberry and blackberry.
Berries available about July to August. Used by Mascouten and Huron; dried and stored (Skinner, 1926: 285; Raddisson, 1885: 143). Archaeological seeds found at the Goldsmith and Feurt sites in Ohio, Manitoulin Island, Ontario (UMMAEL Rept. 139), and at ten locations in the Juntunen site in Michigan. Found growing at the Wexford County sites.

58. *Rubus canadensis* L. Smooth blackberry, dewberry.
Berries available in August and early September. Used by Ojibwa and Iroquois (Gilmore, 1933: 133; Waugh, 1916: 127). Grows from Tennessee to central Ontario and Minnesota in thickets and clearings. K.

59. *Rubus allegheniensis* Porter. Highbush blackberry, sow-teat blackberry.
Berries available in July to August. Used by Ojibwa, Menomini, Sauk-Fox, and Potawatomi; dried for winter (Gilmore, 1933: 133; Smith, 1923: 71; 1928: 264; 1933: 79). Grows from southern Ohio to northern Lake Superior in dry clearings and thickets; usually abundant. K. Also used for medicine.

60. *Prunus pumila* L. Sand cherry. N.
Berries available in July to August. Used by Ojibwa, Menomini, and Potawatomi (Smith, 1932: 409; 1923: 71, 1933: 107). Grows from southern Ohio to central Ontario in sandy and gravelly shores, thickets, wet soil, rocky situations, and sandy forest openings. K. Archaeological seeds found at three locations in the Juntunen site in Michigan (perhaps the next species instead).

61. *Prunus virginiana* L. Chokecherry. N.
Berries available in July to August. Used by Ojibwa, Menomini, Potawatomi, Sauk-Fox, Mascouten, Winnebago, and Iroquois; (Densmore, 1928: 291; Smith, 1923: 71; 1933: 108; 1928: 264; Skinner, 1926: 284; Radin, 1923: 118; Waugh, 1916: 128); dried for winter. Grows from southern Ohio to James Bay (common throughout Michigan) in thickets, shores, and edges of woods. IR, K. Found growing at the Juntunen site and at the Mikado, Rifle River, and Missaukee enclosures. Also used for medicine.

62. *Prunus pensylvanica* L. Fire cherry, pin cherry, bird cherry. N.

Berries available in July to August. Used by Ojibwa, Potawatomi, and Iroquois (Smith, 1932: 385, 1933: 77; Waugh, 1916: 128); dried for winter; whole brances in fruit collected (resulting in pruning?). Grows from northern Illinois to James Bay in dry woods, recent burns and openings, sand lands, and hillsides. IR, K. Archaeological seeds found at fourteen locations in the Juntunen site in Michigan; seed of some cherry species found at the Alway site in Ontario (Jury, 1937: 2). Also used for medicine.

63. *Podophyllum peltatum* L. May apple, mandrake. S.

Fruit available in July to August. Used by Ojibwa, Menomini, Sauk-Fox, Kickapoo, and Iroquois (Gilmore, 1933: 130; Smith, 1923: 62; 1928: 256; Chamberlain, 1901: 3; Waugh, 1916: 129). Grows from the Gulf of Mexico to Minnesota and southern Ontario (very common in Michigan) in low rich woods and thickets. K.

64. *Vaccinium corymbosum* L. Highbush blueberry, swamp blueberry. S.

Berries available in July to August. Used by Iroquois (Waugh, 1916: 128); Grows from the Gulf of Mexico to Minnesota, southern Ontario, and northern Michigan (common throughout Michigan) in marshes, swamps, thickets, and woods. K.

65. *Vaccinium angustifolium* Ait. Low sweet blueberry. N.

Berries available in July to August but earlier than *V. corymbosum.* Used by Ojibwa, Menomini, Potawatomi, and Iroquois (Smith, 1932: 401; 1923: 66; 1933: 99; Waugh, 1916: 128); dried for winter. This and *V. corymbosum* are the most useful blueberries (Newberry, 1887: 44); apparently this is the most important blueberry in the Upper Great Lakes region and one of the most important berry foods in season and dried for winter; it covers the ground in many sections. Grows from northern Indiana to James Bay in dry rocky or sandy soil, burns and clearings, dry open barrens, peats, and rocks. IR. Also used for medicine.

66. *Vaccinium myrtilloides* Michx. Sour-top blueberry, Canada blueberry. N.

Berries available in July to August. Used by Ojibwa, Potawatomi, and Iroquois (Reagan, 1928: 238; Smith, 1933: 57; Parker, 1910: 96); dried for winters; ". . . important and valued article of food among the northern Indians" (Blair, 1911: 103). Grows from northern Indiana to Hudson Bay in dry plains, clearings, moist woods, and swamps. IR, K. Also used for medicine.

67. *Vaccinium* spp. Blueberry.

Large quantities gathered by Ojibwa (Gilmore, 1933: 139). Much gathered by Algonkins in August and September; dried and stored for time of need (Blair, 1911: 103). Among the richest of fruits (Hinsdale, 1932: 9). Berries borne in terminal clusters, thus easily collected in quantity. (The *Vaccinium* species of the bilberry group are not mentioned in the ethnobotanical sources even though they are sweet and quite edible. It is likely that they were little used because the berries are borne in leaf axils rather than in terminal clusters.) Found archaeologically at the Juntunen site. Found growing at a Wexford County site and at the Missaukee enclosures.

Late Summer

68. *Vaccinium vacillans* Kalm. Dryland blueberry, late low blueberry. S.

Berries available in August to September. Used by Ojibwa (Smith, 1932: 401). Grows from Georgia to southern Ontario and northern Michigan in dry open woods, thickets, and clearings in dry places. IR, K.

69. *Vaccinium oxycoccos* L. Small cranberry. N.

Berries available in August to September and through winter. Used by Ojibwa (important food), Menomini, Potawatomi, and Iroquois (Smith, 1932: 369, 1923: 65, 1933: 99; Waugh, 1916: 128). Grows from northern Ohio and northern Wisconsin to northern Canada in cold sphagnum bogs and boggy or peaty soil. IR. Also used for medicine.

70. *Gaultheria hispidula* (L.) Bigel. Moxie plum, creeping snowberry. N.

Berries available in August and September. Used by Ojibwa (Densmore, 1928: 288). Grows from southern Michigan to Hudson Bay in mossy (often coniferous) woods and knolls and southward in bogs. K. Also used for beverage.

71. *Arctostaphylos uva-ursi* (L.) Spreng. Bearberry. N.

Berries available in July to October and through winter. Used by Ojibwa (Densmore, 1928: 287). Grows from northern Ohio and northern Illinois to northern Canada (common in northern, rare in southern Michigan) in dry sandy or rocky soil and exposed rock and sands. IR. Found archaeologically at four locations in the Juntunen site and found growing there; found archaeologically on Isle Royale. Also used for medicine, smoking, charm, and technology.

72. *Prunus serotina* Ehrh. Black cherry, rum cherry. S.

Berries available in August to September; good crop every year. Used by Ojibwa, Menomini, Potawatomi, and Iroquois (Smith, 1932: 409; 1923: 71, 1933: 108; Waugh, 1916: 128); dried for winter. Grows from the Gulf of Mexico to Lake Superior and southern Ontario (frequent in southern, rare in northern Lower Michigan) in rich, moist soil, dry gravely or sandy soils, dry woods, and fence rows. K. Found growing at the Wexford County sites, at the Juntunen site, and at the Missaukee enclosures. Also used for medicine.

73. *Prunus nigra* Ait. Canada plum.

Fruits available in August to September. Used by Ojibwa, Sauk-Fox, and Iroquois (Smith, 1932: 409; 1928: 263; Waugh, 1916: 128). Grows from northern Ohio and northern Illinois to Lake Superior and southern Ontario (more common to the north) in rich alluvial soil, along streams, in thickets, and along borders of woods. Archaeological seeds found at two locations in the Juntunen site in Michigan, at Dutton's Ash Pit in Ontario (UMMAEL Rept. 313), at the Lawson site in Ontario (Wintemberg, 1939: 6), and at the Serpent Mound in Ontario (UMMAEL Rept. 409). The Ripley site in Chautauqua County, New York (Parker, 1907: 547), the Westfield Site in the same county (Guthe, 1958: 39), the Roebuck site in Ontario (Wintemberg, 1936: 12), the Sidey-Mackay site in Ontario (Wintemberg, 1946: 155), the Alway site in Ontario (Jury, 1937: 2), the Middleport site in Ontario (Wintemberg, 1948: 3), the Moccasin Bluff site in Berrien County, Michigan (Hale G. Smith, Field Notes, UMMA), and the Verchave II site (20 MB 181) in Macomb County, Michigan, produced plum pits of this or the following species. Also used in dyeing.

74. *Prunus americana* Marsh. Wild plum. S.

Fruits available in August to October (usually a good crop annually). Used by Ojibwa, Sauk-Fox, and Iroquois (Densmore, 1928: 291; Smith, 1928: 263; Waugh, 1916: 128); dried for winter. Grows from the Gulf of Mexico to New York, northern Lake Erie, northern Lower Michigan, western Upper Michigan, and northern Minnesota in thickets and borders of streams, swamps, and woods. K. Found archaeologically at Dutton's Ash Pit in Ontario (UMMAEL Rept. 313), at the McGraw Hopewell site in Ohio (Yarnell, 1958) at the Baum, Gartner, Feurt, Baldwin, Cramer, Merion, and Franks sites in Ohio, and at the Verchave II site (20 MB 181) in Macomb County, Michigan. Also used for medicine.

75. *Pyrus melanocarpa* (Michx.) Willd. Black chokeberry.
Berries available in August to October. Used by Potawatomi (Smith, 1933: 107). Grows from Tennessee to central Ontario in swamps and low grounds. IR, K. Also used for medicine.

76. *Corylus americana* Walt. Hazelnut. S.
Nuts available in August to September (good crop every two or three years). Used by Ojibwa, Menomini, Kickapoo, and Iroquois (Smith, 1932: 397, 1923: 63; Hunter, 1823*b*: 265; Waugh, 1916: 123); stored for winter. Grows from Georgia to northern Lower Michigan in thickets; very common. K. Found archaeologically at the Roebuck site in Ontario (Wintemberg, 1936: 12), at the Boyer's Run Rock Shelter (36 VE 6) in Pennsylvania (Thompson, 1954: 118), at the Baum, Cramer, and Gartner sites and Ash, Kettle Hill, and Canter's Caves in Ohio, and this or *C. cornuta* at the Verchave II site (20 MB 181) in Macomb County, Michigan. Also used for medicine, dye, and technology.

77. *Corylus cornuta* Marsh. Beaked hazelnut. N.
Nuts available in August to September (good crop about every five years). Used by Ojibwa, Potawatomi, and Iroquois (Smith, 1932: 398; 1933: 97; Waugh, 1916: 123). Grows from southern Michigan and southern Wisconsin to James Bay (rare in southern, common in northern Michigan) in thickets, along stream banks, and on borders of woods. IR. Found archaeologically at 25 locations at the Juntunen site and at Indian Point, Isle Royale (UMMA collections). Found growing at the Juntunen site and a Rifle River enclosure. Also used for medicine.

78. *Nemopanthus mucronata* (L.) Trel. Mountain holly, catberry. N.
Berries available in August to September. Used by Potawatomi (Smith, 1933: 39). Grows from northern Ohio, northeastern Illinois, and northeastern Wisconsin to central Ontario (frequent throughout Michigan) in deep cedar and tamarack bogs, damp woods, thickets, and swamps. K. Also used for medicine.

79. *Cornus canadensis* L. Bunchberry, dwarf cornel. N.
Berries available in August remaining on till October. Used by Ojibwa and Potawatomi (Densmore, 1928: 307; Smith, 1933: 98). Grows from northern Ohio and northern Illinois to central Hudson Bay in woods, thickets, and damp openings. IR, K. Also used for medicine.

80. *Smilacina racemosa* (L.) Desf. False Solomon's seal, false spikenard.

Berries available in August to October. Used by Mascouten (Smith, 1933: 63); root used by Ojibwa (Smith, 1932: 407). Grows from Alabama to southern Ontario and throughout the Great Lakes region in woods, clearings, and bluffs. IR, K. Found growing at the Mikado, Rifle River, and Missaukee enclosures. Also used for medicine.

81. *Lathyrus palustris* L. Marsh vetchling. N.

Seeds available in late summer? Used by Ojibwa and Iroquois (Gilmore, 1933: 133; Parker, 1910: 93). Grows from southern Ohio and southern Missouri to central Hudson Bay in damp thickets, shores, meadows, and marshes. IR, K. One beach pea, *Lathyrus japonicus*, was present in the 250-year-old Pic River site, near Heron Bay, on the north shore of Lake Superior (Quimby, 1961: 83).

82. *Lathyrus venosus* Muhl. Vetchling, wild pea.

Seeds available in late summer? Used by Ojibwa (Densmore, 1928: 290). Grows from Arkansas to James Bay in rich woods, thickets, stream banks, and dry or sandy soils. K. Also used for medicine and charm.

83. *Viburnum edule* (Michx.) Raf. Squashberry, mooseberry, *pimbina*. N.

Berries available in August to September and lasting over winter. Used by Ojibwa (Densmore, 1928: 294). Grows from northern Minnesota, northern Wisconsin, northern Michigan, and northern New York to Labrador and Alaska in woods and thickets. IR.

84. *Viburnum lentago* L. Sheepberry, nannyberry. N.

Berries available in September and through autumn. Used by Ojibwa, Menomini, and Iroquois (Smith, 1932: 398, 1933: 63; Waugh, 1916: 128). Grows from southern Ohio to James Bay (frequent throughout Michigan) in rich, moist soil at forest borders, streambanks, roadsides, and woods. K. Also used for medicine.

85. *Viburnum trilobum* Marsh. Highbush cranberry, *pimbina*. N.

Berries available in September and lasting over winter. Used by Ojibwa, Menomini, and Iroquois (Gilmore, 1933: 141; Smith, 1923: 63; Waugh, 1916: 128). Grows from northern Indiana to James Bay (common throughout Michigan) in low grounds, stream banks, and swamp borders. IR, K. Found growing at the Juntunen site. Also used for medicine.

86. *Zizania aquatica* L. Wild rice.

Seeds available in September (stalks tied about two weeks earlier to prevent seed dispersal). Used by Ojibwa, Menomini, Potawatomi, Sauk-Fox, Kickapoo, Winnebago, Dakota, Assiniboin, Ottawa, Algonkin, Iroquois, and Huron; stored for winter (Jenks, 1900; Stickney, 1896; Waugh, 1916: 98). Grows from the Gulf of Mexico to central Ontario (common through Michigan but most abundant northwestward from Wisconsin) in marshes, shallow ponds, lake shores, and slow-moving stream borders in mud bottoms. K. Possibly found archaeologically in Kettle Hill Cave in Ohio (UMMAEL Rept. 15) but not elsewhere even though the seeds survive nearly intact after burning.

Early Autumn

87. *Vitis riparia* Michx. Frost grape, riverbank grape (formerly *V. vulpina*).

Fruit available in August to autumn. Used by Ojibwa (after frost), Menomini, and Iroquois (Smith, 1932: 411; 1923: 72; Waugh; 1916: 128); dried for later use. Grows from Tennessee to southern Ontario and Manitoba (throughout Michigan but infrequent in the pine region) in rich thickets, alluvial soil, and stream banks. K. Grape seeds found archaeologically at the Baum site and at Ash, Kettle Hill, and Canter's Caves in Ohio, at the Krieger site in Ontario (UMMAEL Rept. 289), at the Dumaw Creek site in Oceana County (Quimby, personal communication, 1961), the Verchave II site (20 MB 181), the Feeheley site, and the Juntunen site in Michigan. Grapes found growing at a Rifle River enclosure and at a Wexford site. Also used for medicine.

88. *Empetrum nigrum* L. Black crowberry. N.

Berries available in autumn (July to November) and through winter. Used by Ojibwa (Reagan, 1928: 243). Grows from northern New York, Upper Michigan, and northern Minnesota to the Arctic in rocky places and peaty soil. IR. 89.

89. *Sambucus canadensis* L. Common elder.

Berries available in September and October (easily collected in quantity). Used by Ojibwa, Sauk-Fox, and Iroquois (Gilmore, 1933: 142; Smith, 1928: 256; Waugh, 1916: 128). Grows from the Gulf of Mexico to central Ontario (common throughout Michigan) in moist ground, stream banks, and following settlements, often in large numbers in a place. IR, K. Stinking elder seeds found archaeologically at seven locations at the Juntunen site (see no. 30). Common elder also used for medicine.

90. *Crataegus* spp. Hawthorn, thornapple. S.

Fruits available in September and October (thorns for sewing gathered in summer). Used by Ojibwa, Potawatomi, Sauk-Fox, and Iroquois (Smith, 1932: 409; 1933: 107; 1928: 263; Waugh, 1916: 128); dried for winter. Grows north to James Bay in thickets, rocky ground, stream banks, open woods, open ground, and borders of woods. Also used for medicine, smoking, and technology. Of about 25 species growing within the Upper Great Lakes region, 14 occur in the southerly and easterly sections, 10 grow throughout the region, and one grows in the northwesterly section. Found archaeologically at the Verchave II (20 MB 181) site in Macomb County, Michigan. Found growing at the Mikado and Missaukee earthworks and at the Juntunen site; also grows on various other sites in Michigan (E. M. Greenman, personal communication, 1961).

91. *Pyrus coronaria* L. Crab apple. S.

Fruits available about October; good crop every two to four years. Used by Ojibwa, Menomini, and Iroquois (Reagan, 1928: 236; Morse, 1822: 52; Waugh, 1916: 129). Grows from Tennessee to central New York, southern Ontario, central Lower Michigan, Wisconsin, and Minnesota in bottoms, wooded slopes, thickets, and clearings. K.

92. *Asimina triloba* (L.) Dunal. Paw-paw. S.

Fruits available about October. Used by Iroquois (Waugh, 1916: 129). Grows from Florida and Texas to western New York, southern Ontario, southern Lower Michigan, and Illinois (common in the Grand and Maple River valleys) in rich woods and alluvium. K. Seeds found archaeologically at Baum, Gartner, Feurt, and Cramer sites in Ohio. Inner bark fiber found in Ohio rock shelter textiles.

93. *Vaccinium macrocarpon* Ait. Large cranberry, American cranberry. N.

Berries available in September to November and through winter. Used by Ojibwa, Menomini, Mascouten, and Iroquois (Reagan, 1928: 238; Skinner, 1921: 163; 1926: 285; Waugh, 1916: 128); dried and stored. Grows from southern Ohio to central Ontario (common throughout Michigan) in open bogs, swamps, and wet shores. IR.

94. *Castanea dentata* (Marsh.) Borkh. Chestnut. S.

Nuts available in October to November; good crop most years. Used by Iroquois (Waugh, 1916: 123). Grows from Alabama to New York, southern Ontario, southeastern Lower Michigan, and southern Minnesota in well-drained gravelly or rocky soil, glacial drift, and hillsides. Found archaeologically at the Baum and Gartner sites and Kettle Hill and Canter's Caves in Ohio, at the Westfield site in Chautauqua County (Guthe, 1958: 39) and the Bates site (A.D. 1300 ± 200, M-762) in Chanango County, New York (Ritchie, 1959: 11), and at the Holcombe site in Macomb County, Michigan.

95. *Fagus grandifolia* Ehrh. Beech. S.

Nuts available in October to November (however, primarily obtained by collecting chipmunk and deer mouse stores in winter); good crop every two or three years. Used by Ojibwa, Menomini, Potawatomi, and Iroquois (Smith, 1932: 401, 1923: 66; 1933: 100; Waugh, 1916: 123); stored for winter. Grows from the Gulf of Mexico to southern Ontario, eastern Upper Michigan, and eastern Wisconsin (common in Lower Michigan, especially in the northern portion) in deep, rich, well-drained loam. K. Found archaeologically at the Roebuck site in Ontario (Wintemberg, 1936: 12) and at three locations in the Juntunen site. Found growing at the Hubbard Lake and Wexford County sites and at the Mikado, Rifle River, and Missaukee earthworks. Also used for medicine and technology.

96. *Juglans nigra* L. Black walnut. S.

Nuts available in October; good crop irregular. Used by Sauk-Fox and Iroquois (Smith, 1928: 259; Waugh, 1916: 123). Grows from the Gulf of Mexico to southern Ontario, southern Lower Michigan and Minnesota, in rich woods. K. Found archaeologically at the McGraw Hopewell site in Ohio (Yarnell), at the Florence, Baum, Gartner, and Feurt sites and Ash and Kettle Hill caves in Ohio, at the Muntz site in Warren County, Pennsylvania (UMMAEL Rept. 260), at the Westfield site in Chautaugua County, New York (Guthe, 1958: 39), near Aylmer in Ontario (UMMAEL Rept. 176), at Moccasin Bluff in Berrien County, Michigan (UMMA Collections), at the Verchave II site (20 MB 181) in Macomb County, Michigan, at the Feeheley site in Saginaw County, Michigan, and at the Spoonville Hopewell site in Ottawa County, Michigan. Also used in dyeing.

97. *Juglans cinerea* L. Butternut. S.

Nuts available in October; good crop every two or three years. Used by Ojibwa, Menomini, Potawatomi, Sauk-Fox, and Iroquois (Smith, 1932: 405; 1923: 68; 1933: 103; 1928: 259; Waugh, 1916: 123); stored for winter. Grows from Georgia to southern Ontario and northern Lower Michigan (common in the southern half of Lower Michigan) in low, rich woods, riverbanks, and low hillsides. K. Found archaeologically at the Baum, Gartner, Feurt, and South Park sites and Ash, Kettle Hill, and Canter's Caves in Ohio, at the Muntz site in Warren County, Pennsylvania (UMMAEL Rept. 260), at the Ripley (Parker, 1907: 547), Reed Fort or Richmond Mills (Parker, 1918: 29), and Westfield (Guthe, 1958: 39) Iroquoian sites, the Castle Creek (Ritchie, 1944: 64), Bates (Ritchie, 1959: 11), and Miller (W. A. Kenyon, personal communication, 1963) Owasco sites and the Frontenac Island Archaic site (Ritchie, 1945: 8) in New York, at the Uren, Roebuck, Lawson, and Sidey-Mackay Iroquoian sites in Ontario (Wintemberg; 1928: 4; 1936: 12; 1939: 6; 1946: 155), and at the Norton site in Macomb County, Michigan (UMMAEL Rept. 228), 10 locations in the Feeheley site, Saginaw County, and the Spoonville Hopewell site, Ottawa County, Michigan. Also used for medicine and dye.

98. *Carya ovata* (Mill.) K. Koch. Shagbark hickory. S.

Nuts available in October; good crop every one to three years. Used by Ojibwa, Menomini, Potawatomi, Sauk-Fox, and Iroquois (Smith, 1932: 405; 1923: 68; 1933: 103; 1928: 259; Waugh, 1916: 123); stored for winter. Grows from the Gulf of Mexico to southern Ontario, central Lower Peninsula, and central Wisconsin (common in the southern half of the Lower Peninsula) in rich woods, bottoms, and slopes. K. Found archaeologically at the Baum, Gartner, Baldwin, and Schizler sites in Ohio and at the Verchave II site (20 MB 181) in Macomb County, Michigan. Nuts of some species of hickory were found at the McGraw Hopewell site in Ohio (Yarnell), at the Cordray, Dominion, Wolford, Dunlap, Cowan Creek, Florence, Feurt, Cramer, and Merion sites and Ash, Kettle Hill, and Canter's Caves in Ohio, at the Muntz site in Warren County, Pennsylvania (UMMAEL Rept. 260), at the Boyer's Run Rock Shelter (36 VE 6) in Pennsylvania (Thompson, 1954: 118), at the Ripley (Parker, 1907: 547), Reed Fort or Richmond Mills (Parker, 1918: 29), Alhart (Ritchie, 1930: 67), and Westfield (Guthe, 1958: 39) Iroquoian sites, and Owasco Lake and Castle Creek Owasco sites, the Vine Valley site (Ritchie, 1944: 64, 83) and the Frontenac Island Archaic site (Ritchie, 1945: 8) in New York, at the Lawson site in Ontario (Wintemberg, 1939: 6), and at three locations in the Feeheley site in Saginaw County, site 20 MB 33 in Macomb County (UMMAEL Rept. 403), and site 20 MR 5 in Monroe County (UMMA Collections), Michigan. Wood used to make tools and weapons.

99. *Carya cordiformis* (Wang.) K. Koch. Pignut, butternut, swamp hickory. S.

Nuts available in October; good crop every three to five years. Used by Iroquois (Waugh, 1916: 123). Grows from the Gulf of Mexico to southern Ontario, northern Lower Michigan, Wisconsin, and Minnesota (common in the southern half of Lower Michigan) in wet to dry woods, stream banks, and swamps. K. Found archaeologically at the Schizler, Baum, and Gartner sites in Ohio and at the Lawson site in Ontario (Wintemberg, 1939: 6).

Carya ovalis (Wang.) Sarg. and *C. Tomentosa* Nutt. grow northward as far as the southernmost part of the Upper Great Lakes region, the latter, at least, being used by the Indians.

Autumn

100. *Quercus macrocarpa* Michx. Burr oak, mossy cup oak. S.

Acorns available in August and September (later on the ground); good crop every two to three years. Gathered in late autumn by Ojibwa and buried for winter or spring (Densmore, 1928: 320). Grows from Tennessee to Lake Superior and southern Ontario (common throughout Michigan) in bottom lands, rich woods, and fertile slopes. K. Also used for medicine and dye set.

101. *Quercus rubra* L. Red oak. S.

Acorns available in September and October (later on the ground); good crop every two to three years. Used by Ojibwa (important starchy food), Potawatomi (use all kinds of acorns), and Iroquois (Smith, 1932: 369; 1933: 100; Waugh, 1916: 123). Grows from Alabama to Lake Superior and southern Ontario (sparingly north of central Lower Michigan) in dry or upland woods, stream banks, well-drained soils, glacial drift, and rich, moist loam. IR, K. Found growing at the Mikado enclosure. Also used for medicine and dye.

102. *Quercus alba* L. White oak. S.

Acorns available in September and October (later on the ground); good crop every four to ten years. Used by Ojibwa (depended on it for soup stock), Menomini (used all acorns), Sauk-Fox (used all acorns), and Iroquois (Smith, 1932: 401; 1923: 66; 1928: 257; Waugh, 1916: 123). Grows from the Gulf of Mexico to southern Ontario, Upper Michigan, and northern Wisconsin (rare in Upper Michigan; common in Lower Michigan, especially in the southern half) in open exposures of dry woods. K. Also used for medicine.

103. *Quercus velutina* Lam. Black oak. S.

Acorns available in September and October (later on the ground); good crop every two to three years. Supposedly used by Ojibwa (Smith, 1932: 402) but does not grow in their territory of early historic times. Grows from the Gulf of Mexico to the southern tip of Ontario, central Lower Michigan, and central Wisconsin in dry woods, glacial drift, dry or gravelly uplands, and poor soils. K. Also used for medicine and dye.

104. *Quercus* spp. Oaks. S.

Acorns available in September to November (later on the ground). All Algonkians used all available species and removed tannin from acorns by soaking in hot wood ash lye water (Smith, 1932: 402). Six additional species grow northward to southern Lower Michigan. K. Found archaeologically at the Fisher site in northeastern Illinois (UMMAEL Rept. 75), at the McGraw Hopewell site in Ohio (Yarnell), at the Franks, Feurt, and Ash Cave sites in Ohio, at the Richmond Mills (Parker, 1918: 29), Westfield (Guthe, 1958: 39), and Silverheels (Parker, 1910: Pl. 31) Iroquoian sites, the Bates and Castle Creek Owasco sites, the Vine Valley site (Ritchie, 1959: 11; 1944: 64, 187), and the Frontenac Island Archaic site (Ritchie, 1945: 8) in New York, at the Lawson site (Wintemberg, 1939: 6), the Alway site (Jury, 1937: 2), the Sidey-Mackay site (Wintemberg, 1946: 155), and a site near Aylmer (UMMAEL Rept. 176) in Ontario, and at the Moccasin Bluff site (Hale G. Smith, Field Notes, UMMA), five locations at the Feeheley site, the Stroebel site, the Verchave II site (20 MB 181), three locations in the Juntunen site, and the Spoonville Hopewell site in Michigan.

32. *Chenopodium* sp. *(C. album* L.?) Goosefoot, pigweed, lambs-quarters.

Seeds available in autumn to early winter. Used by Ojibwa (Stowe, 1940: 12). Grows throughout the Great Lakes region and beyond in cultivated and waste ground. Archaeological seeds found in the Florence and Cowan Creek Adena Mounds (Goslin, 1957: 44) and Ash Cave and Kettle Hill Cave in Ohio, at the Goessens site in Ontario (UMMAEL Rept. 303), and at six locations at the Juntunen site in Michigan.

105. *Amphicarpa bracteata* (L.) Fern. Hog peanut, ground bean.

Beans available in autumn. Used by Ojibwa (raided rodent stores in autumn) and Sauk-Fox (Smith, 1932: 405; 1928: 259); important for all tribes within its range (Gilmore, 1933: 134); vole nests often contain several quarts of beans (Fernald and Kinsey, 1943: 257). Grows from the Gulf of Mexico to central

Ontario in damp woodlands and thickets. K. Also used for medicine.

9. *Nelumbo lutea* (Willd.) Pers. American lotus. S.
Seeds available in autumn, rootstocks in autumn and spring. Used by Ojibwa, Potawatomi, and Sauk-Fox (Smith, 1932: 407; 1933; 105; 1928: 263). Seeds found archaeologically with Woodland and Iroquoian pottery at the Johnson site on Isle Royale (Quimby, 1939: 222).

106. *Lathyrus ochroleucus* Hook. Creamy vetchling. N.
Rhizomes available in autumn? Used by Ojibwa; stored in pits (Smith, 1932: 372). Grows from central Ohio and central Illinois to central Ontario in dry or moist woods, slopes, and rocky banks. IR, K. Also used for medicine.

107. *Polygonatum* spp. Solomon's seal (three species). S.
Rhizomes available in autumn. Used by Iroquois (Parker, 1910: 105). Grows from the Gulf of Mexico to Lake Superior and southern Ontario in woods and thickets. IR, K. Also used for medicine and charm.

108. *Sagittaria latifolia* Willd. Duck potato, *wapato*, arrowhead. N.
Tubers available in late autumn. Used by Ojibwa, Potawatomi, and Sauk-Fox (Densmore, 1928: 320; Smith, 1933: 95; 1928: 254); taken from muskrat caches when possible; dried and stored. Grows from the Gulf of Mexico to James Bay in water or wet places. IR, K. Also used for medicine.

109. *Sagittaria cuneata* Sheldon. Wapato, arrowhead. N.
Tubers available in late autumn. Used by Ojibwa and Menomini (Smith, 1932: 353; 1923: 61); taken from muskrat and beaver caches; dried for winter. Grows from central Illinois and central Ohio to Hudson Bay in calcareous or muddy shores and shallow water. K.

110. *Symplocarpus foetidus* (L.) Nutt. Skunk cabbage.
Rhizomes available in late autumn. Used by Iroquois (Waugh, 1916: 118). Grows from southern Ohio to southern Ontario and Lake Superior in wet meadows, swampy woods, and thickets. IR, K. Also used for medicine.

111. *Lycopus asper* Greene. Bugleweed, crow potato.
Tubers available in late autumn to spring? Used by Ojibwa (Densmore, 1928: 320); dried and boiled. Grows eastward to Michigan and Illinois in marshes and wet shores.

112. *Celtis occidentalis* L. Hackberry. S.

Berries available in autumn; good crop most years. Used by Sauk-Fox (Smith, 1928: 265). Grows from Alabama to northern Lower Michigan (common in Lower Michigan) in rich, moist, well-drained soil, gravelly and rocky hillsides and riverbanks; very shade tolerant. K. Found archaeologically at the McGraw Hopewell site in Ohio (Yarnell) and at the Cramer site in Ohio. (Goslin, personal communication, 1963).

The season of availability is unknown or is continuous for the following six food plants, but presumably they are used only in time of food shortage which would most likely be in late winter and early spring before the maple sap harvest.

113. *Parthenocissus quinquefolia* (L.) Planche. Woodbine, Virginia creeper. S.

Cambium used by Ojibwa (Densmore, 1928: 307); root is a mythological food (Smith, 1932: 411). Grows from Florida and Mexico to central Lower Michigan and Minnesota *(P. inserta* grows from southern Indiana to James Bay) in woods, thickets, and banks. K. Found growing at a Wexford County site.

114. *Celastrus scandens* L. Climbing bittersweet, waxwork, staff tree.

Cambium is an important famine food for Ojibwa, Menomini, and Potawatomi, used in winter by Ojibwa to make soup (Smith, 1932: 362; 1923: 63; 1933: 97); tender branches eaten boiled. Grows from the Gulf of Mexico to central Ontario (common throughout Michigan) in woods, thickets, and river banks. K. Also used for medicine.

115. *Parmelia physodes* (L.) Ach. Puffed shield lichen.

Used by Potawatomi in soup (Smith, 1933: 68). Grows throughout the northern United States and sparingly southward frequently on trees and occasionally on rocks in the highlands and pine barrens. A variable species.

116. *Sticta amplissima* (Scop.) Rabh. Spreading leather lichen.

Favorite old food of Ojibwa; this and other lichens gathered at any season and dried for future use by Menomini (Smith, 1932: 406; 1923: 60). Grows throughout the eastern United States usually on large trees.

117. *Umbilicaria* spp. Rock tripe.

Used by Algonkins; most of their families would starve without it (Blair, 1911: 102-103). Grows throughout the northern

United States and northward on rocks, common on exposed cliffs and boulders.

118. *Cladonia rangiferina* (L.) Web. Reindeer moss.
Used by Ojibwa (Reagan, 1928: 246). Grows throughout the northern United States (common northward) in soil.

The following twelve food plants are listed separately because they are judged to be of questionable significance for aboriginal subsistence in the Upper Great Lakes region. The entries include scientific and common names, tribes utilizing the food and the plant part used, and other uses. The sequence of the entries is taxonomic and follows *Gray's Manual of Botany* (Fernald, 1950).

1. *Equisetum pratense* Ehrh. Scouring rush, horsetail.
Ojibwa; nut-like tuber (Reagen, 1928: 248). Also used to make medicinal tea.

2. *Lycopodium selago* L. Mountain clubmoss, fir clubmoss.
Ojibwa (Reagan, 1928: 246).

3. *Lilium philadelphicum* L. Wood lily.
Sauk-Fox; bulb (Smith, 1928: 262). Also used for medicine.

4. *Lilium canadense* L. Canadian lily.
Miami; bulb (Stickney, 1896: 312). Also used for medicine.

5. *Smilax herbacea* L. Carrion flower, Jacob's ladder.
Sauk-Fox; berries (Smith, 1928: 262). Also used for medicine

6. *Asarum canadense* L. Wild ginger.
Ojibwa; root (Densmore, 1928: 287). Also used for flavoring and medicine.

7. *Polygala senega* L. Senaca snakeroot.
Menomini; root (Hoffman, 1896: 292). Also used for charm and medicine.

8. *Rhus typhina* L. Staghorn sumac.
Potawatomi; berries (Smith, 1933: 38). Also used for beverage, dye, medicine, and smoking material. Seeds of this or *R. glabra* found archaeologically at the Juntunen site on Bois Blanc Island, Michigan.

9. *Sium suave* Walt. Water parsnip.
Chipewyan and Cree; root (Fernald and Kinsey, 1943: 291). Also used for charm.

10. *Pycnanthemum virginianum* (L.) Durand and Jackson. Mountain mint, basil.
    Ojibwa; flowers and flower buds (Densmore, 1928: 307). Also used for medicine.

11. *Pedicularis canadensis* L. Lousewort, wood betony.
    Iroquois; greens (Waugh, 1916: 118). Also used for love charm and medicine.

12. *Mitchella repens* L. Partridge berry.
    Iroquois; berries (Waugh, 1916: 128). Also used for medicine and smoking material.

The plant foods can be classified into eight general categories with a tendency for the members in each category to be gathered at about the same season. This is partly because of availability but probably also because of expediency. The categories are as follows: sap and cambium foods (6 sap, 6 cambium), bulbs and tubers (3 bulbs, 14 tubers), greens (11 kinds), flowers and flower buds (3 species), fleshy fruits and berries (51 kinds), seeds (5 kinds), nuts (13 kinds), and lichens (4 species).

There are 118 entries in all, 8 of which name genera only. Three genus entries have no species entries; the other 5 are entered for special purposes. Three plants involve 2 significant products. Thus, in all, there are 116 plant food products listed, though some of them (e.g., various blueberries) are quite similar; but in such cases habitat, range, and/or season of availability are different.

Altogether 69 genera are represented, but several genera include more than one recognized food. For instance, *Ribes* includes currants and gooseberries, *Rubus* includes blackberries, raspberries, etc., *Vaccinium* includes blueberries and cranberries, and *Prunus* includes cherries and plums. The list includes about 80 different recognized foods from the point of view of our own society.

The list does not include beverage and seasoning plants which are listed separately and more briefly in the appendices as are the utility, dye, smoking, charm, and medicinal plants. The majority of plants utilized by the Indians were for medicinal purposes.

The use categories are not sharply defined because they tend to overlap. There is a gradational continuum from smoking to ceremonial and charm to medicinal to beverage and seasoning to food uses. This continuum, however, does not include the technology and dye plants.

Forty-eight food plants have no other uses recorded in the ethnobotanical literature (excluding literature on the Iroquois). The remaining food plants include 114 additional uses distributed as follows: 73 medicinal, 15 utility, 8 each charm and dye, and 5 each smoking and beverage—flavoring. These figures include 12 additional food plants which are appended to the main list because they are judged to have been of questionable significance to subsistence; but the 12 appended plants are not included in any other quantities mentioned in this discussion of food plants.

Native plant foods were available to the Indians throughout the year, though they are most abundant during the summer and fall and available only to a very minor extent during the winter. The food products of most plants are available only at certain seasons, that is, most of the nuts are available only in the fall, while most of the berries are produced during the summer months. The order of plants in the list is, as nearly as possible, the order in which they were exploited through the annual subsistence cycle, beginning in early spring.

Of the 116 plant food products we are considering here, 29 are available and exploited primarily in the spring, 53 in the summer, 27 in the fall, and 7 probably only during late winter, 4 of which are lichens, 2 are cambium foods, and 1 is a berry (Table XII). Many of these plant products are available during more than one season, but probably very few of them were significantly exploited more than once a year. Of the 17 bulbs and tubers collected for food, 8 are reported to be available in fall and spring. It is likely that they are available most of the time, but that fall and spring are the seasons during which gathering them is most expedient and profitable. Four tubers are among the 20 plants foods available to some extent during the winter, but it is unlikely that they were collected when the ground was frozen. Nine of these foods are berries, 4 are lichens, and 3 are sap—cambium foods.

About 40 of the plant foods in the list are said to have been dried and stored for later use, especially during the winter. They include 23 fruits and berries, 6 bulbs and tubers, 5 nuts, wild rice, milkweed flowers, a tree lichen, and maple sugar. Five of these foods are available in the winter including 3 tubers, a cranberry, and a tree lichen. The fruits and berries that were stored include strawberries, juneberries, currants, gooseberries, raspberries, blackberries, cherries, blueberries, plums, grapes, thornapples, and cranberries in order of seasonal availability. The nuts are hazelnut, beechnut, butternut, hickory nut, and acorn. The bulbs and tubers are onion, *wonkapin (Nelumbo)*, *pin* or groundnut *(Apios)*, 2 species of *wapato* or duck potato *(Sagittaria)*, and crow potato *(Lycopus)*.

Of the 30 spring food plants, 11 are southerly and 5 are northerly in distribution. Among the 53 summer food plants there are still 11 of southerly range, but those of northerly distribution number 27. In autumn the number of northern food plants declines to 4, but the southern food plants increase to 14 of 27. The number of southerly spring food plants is primarily due to the prevalence of bulb- and tuber-bearing species at that season. Southerly plant foods are numerous in autumn because of the abundance of nuts at that time. The high number of northerly summer food plants is due to the large number of berries available from July to September. All 8 of the northerly plant foods available in winter are overwintering berries.

Of 112 flowering plants and ferns used for food in the Upper Great Lakes region, 106 are present at the latitude of southern Lower Michigan. In the central Lower Peninsula this number drops to 102 and in the northern part to 95. In the Upper Peninsula it is 88 to Lake Superior where it is 83. At the latitude of northern Lake Superior the number of food plants is 69 and in central Ontario 60. At James Bay it drops to 37 plants, at southern Hudson Bay to 20, and in northern Canada to 12, 9 of which produce edible berries, 2 furnish greens, and 1, a vetchling, is a source of edible seeds. It is not unlikely that a few food plants used in the subarctic north of Lake Suprior were not used or were not available in the Upper Great Lakes region. For instance, Rogers (1962: c47, c52) reports the use of skunk currant *(Ribes glandulosum* Graner) and bilberry *(Vaccinium uliginosum* L.) for food by the Round Lake Ojibwa of northern Ontario. This does not, however, significantly alter the picture of rapidly decreasing numbers of food plants as one moves northward. When these numbers are plotted against latitude, they form a shallow *S*, as if to form part of one side of a bell-shaped curve, the extremes of which would be in the Arctic and Antarctic with the apex at the equator. This is to be expected because the absolute number of species present is greatest in the tropics and smallest in the polar regions. This should have played a considerable part in the differences in subsistence base between the southern and northern parts of the Upper Great Lakes region and in the even greater differences between the subarctic and the Eastern Woodlands.

Not only does the number of native economic plants decrease toward the north, but also the number of cultivated plants and the intensity of agriculture. All three of these phenomena are directly or indirectly resultant from lower temperatures and shorter frost-free seasons. This northward diminution of available vegetable foods would result in low population density or a greater dependence upon animal foods or both. A. B. Stout (1914: 50) points out that "In the northern parts of North America, food plants are rare and the Eskimo was compelled to subsist largely on animal food. We may say that the food of the Indians inhabiting the northern half of North America was three quarters animal. On the other hand, the food of the Indians of southern United States was three quarters vegetable."

There is considerable variation in geographical distribution from one species to the next, though most of the economic plants of the Upper Great Lakes region have ranges that overlap to some extent. Many species with similar ranges, however, are seldom found growing near each other because of differences in habitat preference. Thus, the number of plant products which can be collected at the same place is never very large at any one time. For this reason, at least, attention can be concentrated on one or, at best, only a few plant species at any given time and place. Yet many localities encompass several different habitats within a relatively small area so that at certain seasons a variety of plant products can be collected without traveling any great distance. Such diversity in the collection of plant products, however, is limited to some extent by the necessity of utilizing a wide variety of techniques for exploitation. For instance, the tools and activities expedient for digging tubers are hardly compatible with those for gathering nuts or berries. Furthermore, burning off the litter and ground cover aids in nut gathering but is detrimental to berrying or locating tubers, and there is some overlapping in the seasons of availability of these foods.

There is great disparity between the importance to subsistence of one plant food and another for various reasons. One plant food may be rather infrequent through its range or parts of it. Another may be low in food value. Others may be productive at a season when other economic pursuits have priority or when food is plentiful. Some plant foods might be important for inverse reasons; and if any of these factors were modified sufficiently, there could be a change in the relationship of certain plants to Indian subsistence patterns. Thus it would be easy to underestimate or overestimate the importance of a plant food to an Indian population when many of the factors concerning their relationships to the natural environment were misrepresented or incompletely understood, as in the present situation. The following is an attempt to estimate the approximate relative aboriginal importance of some of the plant foods in the list within the Upper Great Lakes region in general. This is especially hazardous when one considers the ecological diversity existing within the area and our very scanty knowledge of aboriginal conditions.

Perhaps the two most important wild plant foods in early historic times were maple sugar and wild rice. It is suspected that the prehistoric utilization of these foods was not so great because copper kettles were not available for boiling down sap and because wild rice has not been found archaeologically. On the other hand, utilization of the acorn may have been somewhat greater in prehistoric than in historic times as indicated by its importance as a famine food among the agriculturalists. These judgments and those that follow are based primarily upon the number of tribes reported to have utilized the product, the number of reports of use, archaeological occurrences, and statements and indications of importance in the literature.

The most important of the subterranean plant foods was the *pin* or groundnut *(Apios tuberosa)*, at least in the southern part of the region. To the north the *wapato (Sagittaria)* may have been more important, but they probably were not in competition because the Ojibwa gathered the groundnut in spring, and the *wapato* season is late autumn. It is not clear whether this was the digging season or the time in which muskrat and beaver caches were raided. Onions were important and available in fall and spring. The ground bean may have had considerable importance in some localities.

Greens appear to have been an important food in late spring and early summer, but no particular species seems to have been utilized extensively. Probably those which were most available in various localities were used the most. This may often have been milkweed since it often grows profusely in disturbed habitats; but this plant is more often reported as a source of flowers, for which it apparently was relatively important.

Butternuts, hickory nuts, and beechnuts enjoyed considerable popularity within their ranges; but, after acorns, hazelnuts seem to have been the most important of the nut foods. This may have been because they ripen before the acorn season and because they grow on bushes rather than trees, can be picked rather than collected from the ground, and occur somewhat farther to the north. Thus, they are more like berries from the point of view of exploitation, and probably were collected compatibly with berries.

About the only other important foods are berries which are available predominately during the summer, starting with strawberries and the juneberries (in July), huckleberries, currants, gooseberries, raspberries, blackberries, cherries, May apples, blueberries, cranberries, bearberries, plums, and grapes, elderberries, and thornapples in early autumn. Of these, blueberries and blackberries seem somewhat more important. Wild strawberries are popular because of their flavor and because they are early, but they are rather low in food value.

Toward the north, berries probably were more important relative to other plant foods than they were toward the south, because 27 of the 38 most northerly distributed food plants are berries, whereas fruits and berries account for only 12 of the 38 most southerly distributed food plants. Of the remaining 37 plants, another 12 produce edible fruits and berries. On the other hand, the tubers tend to be southerly; and, with the exception of the beaked hazelnut *(Corylus cornuta)*, all of the nuts are southerly in distribution (Table XI).

The following seasonal sequence is the indicated cycle of exploitation of the more important food plants: maple sap in early spring, tubers in spring, greens in late spring to early summer, berries in summer to early autumn, wild rice in late summer, nuts in autumn, tubers in late autumn (see Table XII). In the southern half of the Upper Great Lakes region the agricultural sequence of land preparation, planting, cultivating, and harvesting should be added for a more accurate interpretation of the annual cycle of aboriginal plant--man relationships, at least during the Late Woodland period. Furthermore, the annual sequence of exploitation of animals for food must be integrated with the plant food cycle for a total view of the annual subsistence cycle. This is true especially in the northern part of the area where animal foods were the major subsistence factor.

Nonfood Utilization of Native Plants

About 325 native plant species were utilized by the Indians of the Upper Great Lakes region for purposes that were not essentially food uses. Eighty-two of these plants were also used for food; and 18 were used for beverage and for seasoning foods. Twenty-five plants were used in dyeing, 27 in smoking, 31 as charms and ceremonial objects, 52 for various utility or technological purposes, and 275 in medicine. Thus there were about 510 uses for the 325 plants involved.

These plants may be found listed in the appendices by category of use. The plants listed in each category are arranged in

TABLE XI

NORTH-SOUTH RANGE OF NATIVE FOOD PLANTS

Southern Limit	Northern Limit	Southern Lower Michigan	Central Lower Michigan	Northern Lower Michigan	Upper Michigan	Lake Superior	Northern Lake Superior	Central Ontario	James Bay	Hudson Bay	Northern Canada
Southern Alabama	...	9TN 10T 92B 96N	30T 98N 103N 113C	3C 25G 56B 74B 99N	37F 95N 102N	2C 21T 63B 64B 72B 107T	13G 16C 22T	12G 19T 86S 89B 105S 114C	38F 90B 108T	8T 34B	...
Northern Alabama	...	94N	11T	76N 97N 112B	40B	14C 101N	1C 68B 80B	23G 32GS 44B	26GF	...	...
Tennessee	...	...	6T 91B	50B 54B	28G	100N	15C 87B	17C 58B 75B	82S	...	...
Kentucky	...	...	...	...	...	7T	...	5B 51B	...	...	...

TABLE XI *(CONT.)*

Southern Ohio	. . . .	. . . .	. . . .	. . . .	. . . .	20T 110T 111T	59B	33G 41B 47B 60B 93B	35B 36B 61B 84B . . .	31C 39B	24G 27G 53B 81S . . .
Northern Ohio	. . . .	. . . .	. . . .	. . . .	. . . .	73B	. . . .	4C 78B 106T . . .	42B 62B 65B 85B	66B 109T	55B 69B 71B 79B
Southern Lower Michigan	. . . .	. . . .	. . . .	. . . .	. . . .	. . . .	. . . .	. . . .	18C 29G 77N	70B	. . . .
Central Lower Michigan	. . . . . .	. . . . . .	. . . . . .	. . . . . .	. . . . . .	. . . . . .	. . . . . .	. . . . . .	. . . . . .	. . . . . .	45B 46B
Northern Lower Michigan	. . .	. . .	. . .	. . .	. . .	. . .	43B	. . .	. . .	. . .	. . .
Upper Michigan	. . . . . .	. . . . . .	. . . . . .	. . . . . .	. . . . . .	. . . . . .	. . . . . .	. . . . . .	. . . . . .	49B . . .	83B 88B
Cumulative Totals	. . .	106	102	96	89	84	70	60	37	20	12

Key: B - berries, fruits; T - tubers, rhizomes, bulbs; N - nuts; C - cambium, sap, twigs; G - greens; S - seeds; F - flowers, buds. Numbers refer to numbers in the food plant list.

TABLE XII

FREQUENCY OF PLANT PARTS USED FOR FOOD BY SEASON

Plant Part	Late Winter?	Early Spring	Spring	Late Spring	Early Summer	Summer	Late Summer	Early Fall	Fall	Total
Lichens	4	..	..	..	..	..	..	..	..	4
Cambium	2	4	4	1	1	..	..	..	..	12
Tubers........	..	3	7	1	..	..	..	..	6	17
Greens........	..	..	2	7	2	..	..	..	..	11
Flowers	..	..	..	..	3	..	..	..	..	3
Berries	1	..	..	..	3	26	13	7	1	51
Seeds.........	..	..	..	..	..	..	3	..	1	4
Nuts	..	..	..	..	..	..	2	6	6	14
Totals	7	7	13	9	9	26	18	13	14	116

taxonomic order. Each entry includes pertinent information about the plant species with beverage, flavoring, and utility plants receiving more attention than the others. The literature from which the data were taken is cited in the section on food plants with the exception of a note by Tantaquidgeon (1932) on Montagnais plant uses.

The remainder of this section is devoted to discussion of the various categories of uses of plants other than for food.

Plants Used for Beverage and Flavoring

The sixteen beverage plants listed in Appendix I include perhaps only a small percentage of those formerly used in preparing drinks by the aboriginal inhabitants of the Upper Great Lakes region, but they are probably among the ones most used and most significant to subsistence. Very little food value is evident in most of these beverages, but it is likely that they furnished certain essential vitamins and minerals to a diet which otherwise might be somewhat deficient (McClendon, 1921), especially during the winter months when meat was the primary staple. Thus, the beverage plants available in winter would likely be those of most value to subsistence.

Cartier describes "the pestilence," or scurvy, which afflicted the Indians and his own men at Stadacona (Quebec) during the winter of 1535-36. He had lost twenty five men when he met an Indian who had recovered as a result of drinking a tea made from the leaves of a tree, probably hemlock. Cartier's sick crew members were able to make a rapid and complete recovery as a result of this discovery (Bakeless, 1950: 115-16).

Only 3 of the beverage plants had no other use, 11 were used in medicine, and there were 14 other uses for this group of plants. Leaves were used to make the beverage in 10 cases, berries in 3 cases, and roots in 2 cases.

Since 4 of the 6 flavoring plants are also beverage plants they are all included to make a total of 18 plants in the list. All 6 of the flavoring plants were used for medicine and 3 are recorded as having been used for food.

The 16 plants reported to have been used for the brewing of beverages are hemlock, sweet fern, jack oak (acorns), red mulberry (berries), smooth sumac (berries), panicled dogwood (bark), labrador tea, bog rosemary, leather leaf, wintergreen, creeping snowberry, spice bush, sassafras, witch hazel, staghorn sumac (berries), and mint. The 6 plants reported to have been used for flavoring are wild ginger, spice bush, sassafras, wintergreen, mountain mint, and mint.

Plants Used Medicinally

In addition to the beverages which apparently were consumed primarily for pleasure, to relieve thirst, and, at least in some cases for nutrient and medicinal values, there are 68 plants whose products were utilized in the brewing of teas, ostensibly for medicinal purposes only. Presumably these beverages were less often consumed and were less significant to the aboriginal diet than were those of the preceding group.

Thirteen of the plants in the beverage list are also classified as medicines and for the most part may be classified as medicinal teas, but they are not included in the list of medicinal beverages so that there be no overlapping between the two "drink" categories. They are, however, included with "other medicines." Thus there are 84 plants recorded as "drink plants": 16 beverages and 68 medicinal teas. It is probable, however, that this number should be closer to 100 because a number of the "other medicines" likely were taken as medicinal teas. In addition to this, much, if not most, of the maple and other saps used aboriginally may have been consumed in liquid form rather than as sugar or thick syrup.

None of the plants listed as medicinal beverages is included with the "other medicines" plants which number 207. Thus there is a total of 275 medicinal plants listed in Appendices IIA and B.

Though there are 149 medicinal plants for which no other uses have been reported, the remaining 126 medicinal plants include 171 other uses for a total of 298 uses as compared with 113 uses for the 97 plants used only for nonmedicinal purposes. The medicinal plants include 60 per cent of all other uses including all six of the flavoring plants, 69 per cent of the beverage plants, 76 per cent of the dye plants, 74 per cent of the smoking plants, 55 per cent of the charm plants, 56 per cent of the utility plants, and 57 per cent of the food plants.

Densmore (1929: 127) reports that Ojibwa medicines were gathered mostly in August.

Plants Used as Charms and for Ceremonial Purposes

The 31 plants listed in Appendix III include those which furnished charms for love and success, charms against snakes, lures for hunting and fishing, medical and ceremonial incense, medicines for malicious magic and supernatural dreams, sacred beads, and sacred bundle components. Perhaps some of the medicinal and smoking plants should be considered in this category also. It is possible that the fishing lures were practical to some extent, but there is no evidence of this.

The charm plants have 36 other uses including 17 medicinal, 8 food, 6 utility, and 5 others. Eleven of these plants have no other uses. More than a third of the charm plants are monocots as opposed to only one-tenth for all other plants utilized.

Plants Used for Smoking

Of the 27 plants listed in Appendix IV as having been used in smoking at least 11 are among the 19 plants used in a formula smoked to attract buck deer when hunting. The odor of the smoke is said to simulate the scent carried in the hoof of the doe. Another 12 plants are listed as smoked and 1 as smoked with tobacco. The flowers of 1 plant were smoked ceremonially and of another to drive away evil spirits causing sickness. One plant in the list was used to make pipestems. Twenty of these plants were used medicinally, and 16 were used for still other purposes.

The part smoked of 24 plants is as follows: bark, 8 plants; flowers, 6 plants; leaves, 4 plants; root, 4 plants; and seeds, 2 plants. Thus the Indians were not limited to the smoking of leaves, which might be one indication that smoking was practiced before the introduction of tobacco. Actually, the conclusion is rather obvious that the occurrence of pipes in archaeological sites does not necessarily indicate that the people who used them smoked tobacco or even had knowledge of the plant. There is no tobacco native to the eastern United States. The species planted by the Eastern Indians in early historic times (*Nicotiana rustica* L.) is native to the central Andes (Goodspeed, 1954). The only archaeological occurrences in the East are from Salts Cave and Newt Kash Hollow, Kentucky, in rather dubious contexts. It was quite shallow in the cave deposits at Newt Kash, at least. Tobacco has been found archaeologically only five or six places in the Southwest and not elsewhere, to my knowledge; but tobacco seeds are quite small and not likely to be found, and the leaves are not likely to be preserved. Yet, from present evidence, no antiquity can be attributed to tobacco in the East. Even if it arrived as early as corn, it would still be antedated by pipes. Thus it would appear that tobacco was merely added to a number of plants already used for smoking, primarily, if not exclusively, for ceremonial and magical purposes. When tobacco was the plant smoked, it generally was only one of the components of a mixture often referred to as "kinnikinnik" (which is still a common name for a few of the plants included in the mixture).

Champlain reports that the Indians near Quebec were cultivating tobacco in 1603 (West, 1934: 54), and the Tobacco Indians (Petun, Tionontati) were named such because of their extensive cultivation of the plant. Allouez reports that in 1670 the Sauk, Fox, Potawatomi, and Winnebago had fields of corn, beans, squash, and tobacco (West, 1934: 55).

Apparently the usual method of growing tobacco was to burn a brush pile and to scatter seeds in the ashes but with little further care to the plants. The fire would destroy all competing plants and seed already on the ground. Any competition would be from pioneering weeds, many of which (e.g., chenopod, Indian hemp, and milkweed) would be considered useful themselves, though not so valuable as the tobacco. Skinner (1926: 287) reports that pumpkins were planted with tobacco by the Mascouten. Skinner (1921: 358) also reports that the Menomini disclaim all knowledge of tobacco cultivation. He states, however, for the Menomini (1921: 66) that "The gathering of roots and herbs for medicinal use is always attended by placing tobacco in the holes from which they were dug, with a song or a prayer in honor of Earth Grandmother, whose hairs they are." If the latter two statements are correct, the Menomini probably acquired their tobacco in trade, perhaps for wild rice.

In 1889, George A. West (1934: 62) found a small patch of wild tobacco *(Nicotiana rustica)* in the Menominee valley, within the city of Milwaukee. Since a large Indian village was located on the adjoining highland, it is possible that this tobacco patch is a relict of former cultivation by the Indians. Fernald (1950: 1260) reports that its present occurrence is rare and sporadic in old fields and waste places from Florida and New Mexico north to Minnesota, southern Ontario, New York, and Massachusetts. Introduction by the Indians may be responsible for its occurrence over most of its range.

The plant materials reported to have been used in smoking are willow bark, persicaria flowers, crowfoot seeds, meadowsweet stems (pine stems), thornapple bark, sumac leaves, musquash root, pimpernel seeds, dogwood bark, bearberry leaves, starflower root, arrowwood bark, goldenrod flowers aster flowers, fleabane flowers, everlasting flowers, and yarrow flowers.

Plants Used in Dyeing

The dye plants, like the smoking plants, charm plants, and, presumably, many of the medicinal plants, would seem to have little if any significance for subsistence except possibly in an indirect way. It may be that some of the plants first utilized for dyeing later came to have more important uses, but it is more likely that the reverse actually took place. Only 3 of the 25 dye plants are reported as having been used only for dyeing. Thus it may be that most of the dyes were discovered as a result of the use of the plants for other purposes. Nineteen dye plants were used for medicine, 8 for food, and 14 for other purposes. Thus the nondye uses of the dye plants are 164 per cent of the dye uses. Only the beverage and flavoring plants have a higher percentage of other uses.

The annotated list of dye plants may be found in Appendix V.

Plants Used for Technological Purposes

Next to the food plants, those used for technological purposes are the most important to primitive subsistence. The most useful of these plants probably were paper birch, northern white cedar (arborvitae), and basswood, all rather common trees throughout most of the Upper Great Lakes region. Birchbark was found archaeologically at many locations throughout the Juntunen site on Bois Blanc Island, Michigan; and charred basswood bark was found at one location. Twenty-four other trees were utilized along with 7 shrubs and 18 herbaceous species.

The plant materials utilized were bark elements in 13 instances. The barks of 5 trees were used for making house coverings. Four trees furnished coarse fiber for making cord, bags, nets, etc. Two kinds of bark were utilized for making mats and bags and one for making boxes and baskets. Birchbark was used for making canoes and various utensils, especially buckets for cooking, food storage, and maple sap collection.

The stems of 7 plants (mostly weedy) provided fiber for fine sewing, fabrics, etc. Six were used for making mats, baskets, and bags; and at least 2 plants furnished stems for basket ribs and 1 for thatch.

The leaves of 2 plants were used to make mats for thatching houses and of 2 others for making mats, baskets, and bags. The leaves of 1 plant were used to line food containers.

At least 9 tree species furnish wood for tools, weapons, etc. and 2 for the manufacture of splint basketry. The roots of 6 trees (mostly conifers) were used for stitching canoe seams and of 1 for making bags. Four trees furnish pitch for sealing canoe seams and leaks. Two trees were favored for wigwam poles, 2 furnished stalks for arrows, and 1 furnished thorns for sewing.

The 52 technology plants were utilized for 58 other purposes, 29 of which were medicinal, 15 for food, and 14 for other purposes. Thirteen plants were not recorded as having other uses.

The annotated list of plants used for technological purposes may be found in Appendix F.

IV

ABORIGINAL INFLUENCES ON DISTRIBUTIONS, HABITATS, AND GENETIC VARIATION OF PLANTS

As a culture adapts to the natural environment in which it is situated certain interrelationships develop. Statements about these interrelationships often are phrased only in terms of the influences of natural environment upon culture. This is an oversimplification, because the influences are reciprocal. It is true that environment does influence culture; but culture also, in its technological aspect, has many significant effects upon environment. These effects in turn have countereffects upon culture.

Robert F. Heizer (1955: 1) has commented that ". . . anthropologists have now accumulated sufficient evidence to argue that at any point in time or space where man has occupied a region he has materially affected the soil, the fauna, the flora, and even the climate, through the intermediacy of that one distinctive human possession which we call culture."

The activities connected with modern civilization have created by far the most extensive of the cultural modifications of natural environment. However, there have been many significant changes effected by less advanced cultures as a result of clearing and burning vegetation, collection of food, firewood, and raw materials, disposal of waste, cultivation of the soil, etc. Alteration of physical and chemical properties of the soil, increase of surface light, evaporation, soil erosion and sedimentation, alteration of plant and animal community composition, and dispersal and modification of plants and animals are some of the results of these activities. These results have variously increased and decreased the productivity of the environment for the cultures involved depending upon the kinds, intensities, and duration of the activities upon the environments in which they were carried out.

The northeastern United States is an area for which there is significant evidence of the influence of Indian activities on their environment. Gordon M. Day (1953: 342-43) has written a well-documented review of this evidence which he summarizes:

> These Indians created sizeable clearings for their villages and fields and probably expanded the clearings as they foraged incessantly for firewood and other necessary materials. Over much of the region, they set

fire to the woods to improve travelling and visibility; to drive or enclose game; and to destroy "vermin". They probably exercized some influence on the forest through their control over the animals they hunted and through planting food and medicinal plants. It is certain that their activities destroyed the forest in some places and it is hardly to be doubted that they modified it over much larger areas. Seasonal migrations and the periodic relocation of villages widened the range of Indian influences, which extended into unexpected localities and supposedly uninhabited region

S. W. Bromley, a plant ecologist, has studied extensively the vegetation of southern New England and reports (1945: 153) that Indian burning favored nut trees at the expense of other timber and perpetuated the blueberry heaths which furnish food for large numbers of game animals as well as for the Indians. Hu Maxwell (1910) has reviewed the evidence for the same class of phenomena resulting from fire and clearing by Indians in Virginia and emphasizes the increased production of nuts, fruits, and game that resulted. He discusses, also, the spread of plant species by the Indians, the results of which are still observable.

Heizer (1955: 12) states that habitation sites "have undoubtedly had a tremendous local effect upon soils, plant cover and the like" and that "Many uncultivated plant distributions are best explained as due to having been spread by man." He also points out that these two factors may have been responsible for the frequent occurrence of economic plants on archaeological sites, which in some cases are outside of the natural ranges of the plants involved. Especially striking instances of the divergent vegetation of archaeological sites have been reported for shell mounds in Florida (Small, 1927 and 1933), for archaeological sites in California (Heizer, 1955), for Pueblo ruins in New Mexico (Yarnell, 1958), and for former Aleut village sites (Bank, 1953).

Such phenomena have rarely been reported for the Great Lakes region. This is probably due to rapid leaching of the soil, limited amount of disturbance and sewage deposition by the Indians of the region, rapid succession of the vegetation (i.e., rapid return to natural vegetation in disturbed habitats), and intensive modern disturbance. A few cases are on record, however. W. Jury believes that an apparent association of American chestnut groves and Indian village sites in lower Ontario is the result of planting by Indians (Day, 1953). According to Hedrick (1933) the Iroquois planted the Canada plum, which is found in [western?] New York only near their villages; and they may have planted the Kentucky coffee tree which, in New York, is most commonly found near their village sites. Fenton (1940: 795) writes of old Iroquois village sites which "reveal characteristic floral changes due to human habitation, which the ecologists term residual gardens, on the supposition that many plants were brought there from afield . . . ." Mosely (1931) has suggested that isolated stations of *Opuntia humifusa* Raf., *Actinea herbacea* (Greene) Robins., *Nelumbo lutea* (Willd.) Pers., *Symphoricarpos albus* (L.) Blake, *Gentiana puberula* Michx., and *Ceanothus ovatus* Desf. on the Ohio coast of Lake Erie are due to intentional or unintentional introduction by the Indians. On Walpole Island at the north side of Lake St. Clair, the Indian amateur archaeologists locate archaeological sites by finding heavy stands of sweet grass *(Hierochloë odorata)*, according to Arnold Pilling (personal communication).

In general, the northern stations of *Nelumbo* (pond nuts) are somewhat isolated from the general range of the species to the south. It may be that Indian activities are primarily responsible for the occurrence of these northern stations. The fact that the nuts (acorn-like seeds) of this species were found with Iroquoian pottery on Isle Royale, which is well to the north of its present range, indicates that they were carried great distances by the Indians. The present distribution of *Nelumbo lutea*, however, is somewhat reduced from what it was formerly, and some of the isolated stations may be only relicts of a somewhat wider natural range. Nevertheless, E. D. Merrill (1954: 262-3) unequivocally states that this species is:

> . . . an excellent example of a native American food plant that had been deliberately and widely distributed by the American Indians, if a really good example be needed. It is, however, not a "cultivated" species in the strict sense of that term. The "range" of the American *Nelumbo lutea* Pers. now covers an area from Minnesota to Ontario southward; it occurs in Mexico, Honduras (Lake Yojoa), and Cuba; and is reported from the Magdalena River in Columbia . . . the benighted North American Indians distributed it where favorable conditions existed, i.e., in shallow protected fresh water with a muddy bottom, in lakes, slow streams, etc. And it is, incidentally, about the only American food plant native of the United States (other than the sunflower) that, through the activities of man, extended its range south of the Mexican border.

In addition there is a statement by William Tricher that:

> . . . it was well known to the Indians and cultivated by them in the waters of the Tennessee and Cumberland rivers, and was abundant on the tributaries of the Mississippi. It was carried northward and eastward by the Indians, and was established as far east as Connecticut (Bailey, 1935: 2118).

It is quite likely that the Indians of the Upper Great Lakes region extended the ranges of other plants, especially the cultivated weedy species like sunflower and Jerusalem artichoke whose genetic and distributional histories were greatly influenced by human activities (Anderson, 1956). A. B. Stout (1914: 53) rephrases Havard (1895: 101) to say that the Jerusalem artichoke "is hardy, perennial and prolific, qualities which made it one of the valuable cultivated plants of the Indians. Its use spread from the valleys of the Ohio and Mississippi, where it was native, to Canada and the Atlantic coast."

There are several other weedy species which are usually considered native to the Upper Great Lakes region and which were extensively utilized by the aboriginal inhabitants. Notable among these are a nettle *(Urtica gracilis* Ait.), spreading dogbane *(Apocynu androsaemifolium* L.), Indian hemp *(Apocynum cannabinum* L.) butterfly weed *(Asclepias tuberosa* L.), and common milkweed *(Asclepias syriaca* L.). Two of these species were used for food and four of them were used for medicine, but all five of them were used as a source of fine, strong fiber to make thread and cordage for sewing, weaving, fish lines, etc. Only two other plants in the region are known to have furnished such fiber, a wood nettle *(Laportea canadensis* (L.) Wedd.) and bog-hemp *(Boehmeria cylindric* (L.) Sw.), both of which are found in rather damp and shady habitats as opposed to the more weedy habitats of the others. The tempting conclusion is that there was a tendency for the Indian women to collect and use fiber from the weedy and probably more easily available species. Since the whole plant, or most of it, was collected when it was mature, seeds would often be brought into camp where they could reproduce in readily available waste ground. Thus transportation from one area to another would probably not be unusual. Such an association with man (or rather, woman) could lead to dispersal to other regions along with trade goods, war parties, or perhaps the marrying out of women. Densmo (1929: 30) reports on the Ojibwa: "A woman who wished to sew would often say to the children, 'Go get me a stalk', and the desire twine would be quickly made. It is said that 'there are lots of thread plants at deserted villages'."

Hunt (1940: 43, 56, 59) mentions that "hemp" *(Apocynum?)* was "cultivated" by the Huron and Petun, though this may have amounted only to encouragement of volunteer weeds in their fields and in other disturbed habitats in or near their villages. Gilmore (1933: 121) states that "there was some conscious dissemination and transplanting of useful and desirable species for the sake of convenience." This may have been the means by which some of the fiber plants were disseminated.

Other possible candidates for dispersal by the Indians are a wild onion *(Allium canadense* L.), a chenopod, *(Chenopodium album* L.), a mint *(Mentha arvensis* L.), cleavers or bedstraw *(Galium* spp.), and the wild balsam apple *(Echinocystis lobata* [Michx.] T & G). Seeds of *Galium* and of what strongly appears to be *Chenopodium album* (supposedly a European introduction) were found to be abundant in the Juntunen site on Bois Blanc Island, Michigan. In addition, the prairie sagewort *(Artemisia frigida* Willd.) is reported to have been intentionally planted for medicine by the Potawatomi (Smith, 1933: 49) and used for medicine by the Ojibwa (Densmore, 1928: 287). This is not a native of the Upper Great Lakes region but is adventive eastward from the prairies in dry fields, roadsides, railroads, etc.

During the summer of 1960, I visited many archaeological sites in Michigan and made plant collections at a number of them in the hope of finding plants which might have been introduced by the former occupants of the sites. Most of the plants identified are simply members of the natural vegetation of the areas studied. Many of them are rather weedy, but the disturbed nature of the habitats in which they were growing is mostly the result of lumbering activities and the intense fires which followed. A number of the plants found on sites were used by the Indians. The following lists include only the sites which show definite aboriginal disturbance and which were host to more than three economic plants of the Indians.

Elder, bearberry, chokecherry, hazel, highbush cranberry, thornapple, service berry, and milkweed were found growing at the Juncunen site on Bois Blanc Island. Seeds of the first four were identified from materials excavated there. Growing at the Rifle River earthworks in Ogemaw County were sugar maple, yellow birch, bracken, basswood, white pine, milkweed, aspen, strawberry, raspberry, currant, gooseberry, two-leaved Solomon's-seal, fire cherry, chokecherry, hazel, false Solomon's-seal, beech, oak, grape, and teaberry. At the Mikado (Pine River) earthwork were found red ash, basswood, white pine, aspen, fire cherry, chokecherry, false Solomon's-seal, thornapple, beech, and red oak. The Missaukee earthworks and food pits were host to bracken, basswood, milkweed, aspen, two-leaved Solomon's-seal, blueberry, chokecherry, black cherry, false Solomon's-seal, thornapple, beech, and oak. On a small mound in Kalkaska County, bracken, white pine, aspen, blackberry, and chokecherry were found growing. In Wexford County, a site on the Manistee River was found to support wild leek, bracken, milkweed, aspen, blackberry, gooseberry, blueberry, fire cherry, black cherry, beech, oak, and grape.

None of these plants was found to be growing outside of its natural range and might have been growing on these sites regardless of whether the aboriginal disturbances of the environment had, or had not, taken place at these locations. This is not to say that the Indians had not disturbed the natural vegetational patterns at these sites but that the results of this disturbance are no longer obvious. A possible exception to this is the Juntunen site which is located in the widest part of a long, narrow meadow which parallels the shore but is separated from the lake by a narrow strip of forest. The antiquity of this meadow is not known, but there is a buried indication of podsolization which is below some of the archaeological remains. This raises the possibility that the Indians were originally responsible for the formation of the clearing which has since become only partly reforested. Nevertheless, recent cultivation and settlement is likely to have contributed to its perpetuation. Some of the grasses present are of Old World origin, and they form a thick sod which could be replaced naturally only by slow encroachment of trees from the periphery as a result of shading out the heliophyllous grasses.

It is unfortunate that most of the sites which might furnish significant results in a search for living relicts of plant—Indian relationships have come under the plow or are otherwise more or less destroyed or significantly altered by recent human activities. Of course, many such sites are not discovered until cultivation turns up artifacts. Failure to find relict evidence of such relations in the Upper Great Lakes region may be due to relatively rapid replacement of disclimax vegetation. In addition, few sites in Michigan indicate disturbance extensive enough for prolonged delay of the development of climax vegetation. For instance, an overnight hunting campsite would present minimum conditions for prolonged vegetational differences; whereas an agricultural village site with deep refuse deposits and relatively permanent architectural features would present ideal conditions.

The climax vegetation of a region is normally composed of those species of plants which will continue to reproduce themselves as long as the climate remains the same. If the climate remains constant but the vegetation is changing, however slowly, the climax has not yet been reached. In a forested region the climax vegetation is typically composed of shrubs and herbaceous plants with a high shade tolerance and of trees whose seedlings require, or at least tolerate, considerable shade. If the forest is opened by cutting or burning of the overstory, plant species with a low tolerance for light or a relatively high humidity requirement are at a great disadvantage to those which do well in the absence of shade. The greater the disturbance of the climax plant community, the longer and more complex will be the developmental sequence, or succession, back to the eventual climax condition of the forest.

Many of the more important economic plants, and all of the cultivated plants, utilized by the forest Indians do not occur or are not very productive of economic products under climax forest conditions. Therefore it is often to the advantage of such peoples to disrupt the climax condition of the forest.

About thirty-five of the plants in our list of Indian foods commonly grow on burned sites, clearings and open areas, and along the borders of woods. Another four plants grow in open woods. This, of course, does not include the aquatic or semi-aquatic species, most of which also grow in the open; nor does it include the large tree species that make up the forest, many of which are of limited economic significance. These plants would, in general, be favored by Indian activities which created openings or otherwise increased light or decreased humidity. Only one of these plants (teaberry) is among those whose food products were exploited in the early spring. The remainder were exploited between late spring and early autumn. This is the season when light is least available in a climax deciduous forest. One plant produces sap, one a bulb, one nuts, two seeds, five greens, and about thirty fruits and berries. They include aspen, an onion, hazel, bracken, 2 vetchlings, 2 milkweeds, chenopod, aster, teaberry, 2 strawberries, two-leaved Solomon's-seal, 3 juneberries, 5 *Rubus* species (blackberry, raspberry, etc.), 3 cherries, 2 plums, 2 gooseberries, a grape, 3 blueberries, bearberry, false Solomon's-seal, sheepberry, elder, apple, and thorn-apple. The most important of these probably were hazel nut, blueberry, blackberry, and milkweed. Thus it is the production of the fruits and berries gathered during the summer and also the game animals that feed on these foods that are most increased by creating clearings.

Presumably the aquatic and other hydrophyllic plants would be little benefited by Indian activities unless the disturbance of the soil created by digging tubers is beneficial, as Sauer (1952) has suggested. The food production of maples and nut trees, however, is increased if competition from other trees is reduced by removal. This allows them to develop greater crown areas, gives them greater access to soil water and nutrients, and stimulates production and reproduction. It has been reported a number of times that the Indians were inclined to spare certain kinds of trees at the expense of others. The exploitation of individual trees with high productivity would probably be more expedient at the aboriginal technological level than exploitation of several trees with higher collective but lower individual productivity. For instance, tapping maple trees would be quite a chore in the absence of steel axes; the higher the productivity of the individual tree, the more expedient would be its exploitation.

Of 110 food plants that can be classified in the following categories, the 39 heliophyllic plants constitute the largest class. There are 32 hydrophyllic plants, 22 trees, and only 17 others, most of which are essentially mesic in habitat preference. Thus the main forest food resources are tree products including sap and nut foods. The hydrophyllic plants are important for tubers and wild rice. Thus forest clearing would appear to be beneficial to the Indians as it increased the production of summer foods, primarily the fruits and berries but also hazel nuts and greens, while forest thinning would increase the production of nuts and maple sap for fall and spring foods respectively.

Light burning in forests would keep down seedlings and thus eventually act to thin the forest somewhat. Heavy burning would open the forest and create habitats for the heliophiles. Light burning was practiced in the forest in the fall to clear off the litter so the nuts could be gathered more efficiently. Heavy burning was practiced to clear land for agriculture and to aid in hunting. Driver and Massey (1957: 188, 226) indicate that the Menomini, Winnebago, Wisconsin Ojibwa, Miami, Neutral, and Huron practiced driving game with fire, and that all of the Upper Great Lakes tribes used burning to clear or fertilize agricultural land.

The best evidence that we have for extensive alteration of plant communities by Indian activities comes from Wisconsin where the Fox Indians lived in a transitional zone between prairie grassland and deciduous forest. According to A. W. Schorger (1937:117) many accounts by early travelers indicate that this country was mostly prairie with small patches of forest occurring here and there. At the present time, however, this region is mostly deciduous forest.[1] Clements and Chaney (1936: 33) believe that annual firing could not have been responsible for such a perpetuation of prairie as apparently existed in the Fox country, but many authorities (e.g., Omer C. Stewart and Carl O. Sauer)

[1] I am indebted to Frank B. Livingstone for this interpretation.

disagree with this position. Curtis and Partch (1948) have shown that prairie plant species cannot compete with forest species in this region under natural conditions but that they encroach upon the forest area when annual firing is practiced. This would indicate that burning, and perhaps other activities of the Indian, were responsible for the former prairie condition of the region. Isolated grassy openings in southeastern Michigan and in other areas along the southwestern side of the Upper Great Lakes region may be relicts of similar aboriginal practices that took place as an aspect of their economic pursuits.

Many plant species are subdivided taxonomically into two or more varieties, and varieties are often constituted by two or more forms. Interbreeding between species produces hybrids. Since Fernald's (1950) varieties, forms, and hybrids are the ones to be considered here, his definitions will be used.

> As here used the term VARIETY. . . refers to *a geographic variety,* a strongly fixed variation of a species with the essential reproductive parts unchanged but showing somewhat constant departures in size of parts, shape of leaves, or modification of the less fundamental parts of the flower, etc. and occupying a somewhat segregated geographic area . . . The FORMS . . . are minor departures in vegetative characters (dwarfing, depressing, etc.) indument, multiplication or loss of parts, change of color, etc., striking variations but without fundamental or constant trend over large areas, occurring here-and-there in the broad range of the species . . . HYBRIDS of two usually distinct species frequently occur and in some very plastic and unstable groups, like many genera of the *Rosaceae,* or in *Oenothera,* they seem just now to be endless. . . Their characteristics are usually in some ways a combination of those of the parents but the plants are often more vigorous and floriferous and their recognition is the result of experience and trained judgement. (Fernald, 1950: vii-viii).

The morphological variations by which varieties, forms, and hybrids are distinguished presumably are reflections of genetic variations which, if adaptive, will persist indefinitely and perhaps even lead eventually to a higher taxonomic status if some sort of isolation develops. Many of the activities of man tend to encourage these conditions and processes, which in many instances, have led to domestication (Anderson, 1949 and 1952).

In order to make some estimate of the possible relationships between aboriginal human activities and plant variation in the Upper Great Lakes region, a comparison was made between the variation in plants utilized by the Indians and the variations in the total flora of the region and the total flora listed in *Gray's*

Manual of Botany. Fernald (1950: lxiv) lists a total of 4425 native species, 1487 native named varieties, and 1121 native forms and named hybrids for the flowering plants and ferns of the central and northeastern United States and adjacent Canada. An analysis of 364 native species utilized by the Upper Great Lakes Indians shows that they include 318 native named varieties and 262 native forms and named hybrids. Comparing the two sets of figures we find that the utilized species are 8.2 per cent of the total listed by Fernald but that the varieties of the utilized species are 21.7 per cent and the forms and named hybrids are 23.4 per cent of Fernald's totals. Thus the number of varieties is 265 per cent and of the forms and named hybrids is 285 per cent of that which should be expected of the utilized plants on the basis of random distribution.

Another comparison that can be made is between the 171 utilized species with named varieties which is 47.0 per cent of the 364 utilized species to the 950 species with named varieties listed by Fernald which is 21.5 per cent of the total. Thus the number of utilized species with varieties is 215 per cent of the expected number.

A third comparison is between the number of utilized species and the number of Upper Great Lakes species with more than one native named variety (excluding the genus *Crataegus* for which we do not know the species utilized). Among the utilized plants there are 84 such species or 23.1 per cent. Among the estimated 2000 Upper Great Lakes species 233 or 11.6 per cent have more than one native named variety. This means that the ratio for the utilized plants is 198 per cent or almost twice that to be expected on the basis of random distribution. Counting the total number of varieties involved in this analysis the number of native varieties of utilized species with more than one named variety is also 198 per cent of the expected number.

The rather obvious conclusion to be drawn from these analyses is that the genetic variation among the plant species utilized by the Upper Great Lakes Indians is about twice as great as the genetic variation in the total flora. The question arises then: is this tendency toward high variation in utilized species a cause or a result of utilization, or perhaps neither? If some of the variation is the result of plant—man relationships it should be expected that those plants with the strongest relationships with man would be those exhibiting the highest degree of variation. It might be expected that these plants would be the ones used for food and utility purposes, but the results are inconclusive. The dye plants are highest in forms and named hybrids but lowest in varieties while the smoking plants are highest in varieties but lowest in forms and named hybrids. In over-all variety the order from highest to lowest by use category is charm, food, utility, smoking, medicine, dye, and beverage. It might be expected that charm plants would have a lower relative variation and beverage plants a higher relative variation if the incidence of influence of Indian activities on the plants created a tendency toward greater variation, but perhaps this underrates the degree to which the charm plants came under the influence of disturbance created by human activities. The relative degrees of variation in charm and beverage plants, however, seem even more a paradox when it is recalled that the monocots contributed 35 per cent of the charm plants and none of the beverage plants and that, of the four groups of higher plants dealt with in *Gray's Manual*, the monocots exhibit the smallest incidence of variation.

Another explanation for the relatively high degree of variation in the species utilized is that the Indians tended to select the more variable species for subsistence and other purposes. For instance, many members of the family Rosaceae furnished foods and medicines. Yet, this highly variable family is represented by a smaller percentage of species among the plants utilized than it is among the total flora. It is likely, however, that many species of the family Rosaceae were utilized, especially for food, but are not reported to have been used. This probably is true for the genera *Amelanchier* (juneberries), *Crataegus* (thornapples), and *Rubus* (blackberries and raspberries). In fact, it is not unlikely that Indian activities had a lot to do with the present proliferation of species in these particular genera. The genera *Rubus* and *Crataegus* almost defy conventional classification because of the extensive intergradation occurring between many of the component species. Much of this intergradation may be the result of introgressive hybridization with survival of hybrids in hybrid habitats created by the Indians (Anderson, 1949). The extensive hybridization that occurs among the oaks may be partly the result of the same process. This speculation is supported by knowledge of the extensive utilization of acorns for food by the Indians.

Those families with at least 25 native species listed in *Gray's Manual* and of which more than 15 per cent and no less than 8 species were utilized are the families Pinaceae, Caprifoliaceae, Corylaceae, Fagaceae, Ericaceae, Liliaceae, Umbelliferae, and Ranunculaceae. Three of these families—Corylaceae, Fagaceae, and Ericaceae—have form-hybrid percentages that are more than double the expected figures; and the family Fagaceae is 8 times the expected, because of the large number of oak hybrids. But the family Umbelliferae has only 36 per cent of the expected form—hybrid percentage. Five of these families have less than the expected variety percentage, and none is nearly double the expected figure. Thus we cannot cite heavy utilization of families with high variation to account for the high variation tendency in utilized species except perhaps in the cases of the Corylaceae and Fagaceae. But these two families include a total of only 16 species in the list of utilized plants. They are all trees or large shrubs which have longer life cycles than herbaceous plants and thus would take much longer to modify genetically.

Fernald specifies that a variety occupies "a somewhat segregated geographic area" but that forms are "without fundamental or constant trend over large areas, occurring here-and-there in the broad range of the species." Since the latitudinal ranges of the food plants had already been plotted it was decided to test species variation against distance of north—south distribution. Latitude indices were assigned so that distance of north—south distribution might range from 1 to 16 latitudinal zones in eastern North America, averaging about 2 degrees apiece (actually the range is 4 to 15). The average distribution index for 41 food plants without varieties or forms listed in *Gray's Manual of Botany* is 8.7. For 26 food plants with forms but without varieties the index is 9.2, and for 51 food plants with varieties the index is 10.6. Thus, the average latitudinal range of species with forms, but not varieties, would appear to be about 6 per cent greater than the average for those with neither, while the average for species with varieties would appear to be about 22 per cent greater. Another way to state it is that the average latitudinal range of food plants with varieties appears to be about 19 per cent greater than that of the average for food plants without varieties. Thus the food plants with greater varietal diversity would more likely be available over a greater part of the Upper Great Lakes region, or vice versa.

The only valid conclusion about the relatively high genetic variation in the utilized plants would seem to be that part of the diversity may be ascribed to a tendency to make use of plants with greater variation and that part is due to the influences of Indian activities in creating greater variation through dispersal into new habitats, the creation of new habitats, encouragement of hybridization and weediness, and semicultivation.

V

HISTORY OF THE ABORIGINAL DISTRIBUTION OF CULTIGENS IN THE MIDWEST AS INDICATED BY ARCHAEOLOGICAL REMAINS

As a result of their various subsistence activities the Indians may have brought a number of local plant species under "cultivation," that is breaking the soil around them and removing competitors. In fact, there may have been some propagation of native plants; but this study is concerned with those plants which were intentionally introduced: those which could not be members of the natural flora because their continued existence was dependent upon certain techniques of propagation and cultivation. The members of this category of plants are cultigens.

The earliest site in the eastern United States with cultigen remains that were in close association with radiocarbon dated materials is Newt Kash Hollow in Menifee County, Kentucky. This bluff shelter, dated at 690 ± 150 B.C. and 640 ± 150 B.C., "is a relatively homogeneous site attributed to early Woodland" (Crane, 1956: 665). The site yielded corn, squash, gourd, sunflower, and tobacco (Jones, 1936*b*). Unfortunately, it is not evident which of these plant remains other than sunflower are contemporary with the material from which the date was derived.

There appears to be some confusion regarding the Newt Kash corn. Jones (1936*b*: 148) reports that "only two small fragments of cobs and one ear stalk were found." Webb and Funkhouser (1936: 130) report that "corn cobs were numerous" and illustrate an example which bears little resemblance to any known prehistoric type of cob from the eastern United States, but it does resemble modern cobs such as those which might remain as a result of producing mash. Furthermore, in recapitulating Webb and Funkhouser, Jones (1936 *b*: 162-163) points out that "the lower levels of shelters in Kentucky show no agriculture, agriculture is indicated in the intermediate levels by gourd fragments and hoes, while corn and pottery occur only in the very top levels. The squash seems earlier than corn, and beans do not occur. Tobacco is scarce and probably late."

There are records of cultigens from five other rock shelters in Kentucky and from the vestibule of Mammoth Cave where sunflower seeds were found in the precorn, prepottery level (Nelson, 1917: 53). Gourd was present in Dillard Stamper, Worth Creech,

Little Ash, Red Eye Hollow, and Stephen De Hart Rock Shelters (Webb and Funkhouser, 1936) and in Mammoth Cave (Nelson, 1917: 31, 35, 36). Stephen De Hart Shelter also contained squash (UMMAEL Rept. 47) as did a rock shelter in Harlan County (UMMAEL Rept. 42). All prehistoric squash and pumpkin remains from the eastern United States are *Cucurbita pepo* according to Hugh Cutler (personal communication, 1962), though *C. moschata* "may have spread to the Southeast, perhaps along the Gulf in prehistoric times" (Cutler and Whitaker, 1961: 479).

Webb and Baby (1957: 34) contend that the Newt Kash Shelter is an Adena site. Goslin (1957: 46) lists gourd, squash, and sunflower for Newt Kash in his chapter on Adena food; but corn and beans are conspicuously absent from his list. Webb (1952: 179) states that the Adena people ". . . were the first agriculturalist in this region, having introduced beans, squash, pumpkin, and sunflowers. There is no evidence that they had corn." There appears to be no evidence that they had beans, and Webb fails to mention gourd.

Jones' discussion (1936 *b*: 149-152) of other Newt Kash plant remains and their similarity to those of the Ozark Bluff Shelters argues strongly for the inclusion of *Iva* (marsh elder) and *Chenopodium* (goosefoot) to this postulated early agricultural complex and perhaps *Phalaris* (canary grass) as well. *Iva* especially appears to have been an early cultigen (Black, 1963). *Iva* and *Chenopodium*, in addition to sunflower, were found in direct association with the material from which the Newt Kash dates were derived.

Nelson (1917: 31) reports not only gourd, squash, and sunflower from Salts Cave, an extensive cavern system some three miles northeast of Mammoth Cave, but also corn and tobacco. Nelson gives no indication of the age of these remains but radiocarbon dates of 420 ± 60 B.C. and 280 ± 40 B.C. are reported for Mammoth Cave (Benington and others, 1962: 240). These plants are also reported from Salts Cave by Young (1910: 180, 306, 313-315, 323-328), but in this case, an Old World plant, watermelon, is reported, perhaps mistakenly. This, in addition to the fact that seeds from a Salts Cave gourd germinated and produced plants, casts doubt on extreme antiquity for some of the Salts Cave plant remains. Benington and others (1962) report a radiocarbon date of 1125 ± 140 B.C. for Salts Cave, but they state that the surrounding region ". . . was occupied from the Paleo-Indian period to the Mississippian." Later occupations probably account for the presence of corn, which, like Mississippian pottery, occurs only rather near the cave entrance. Corn was reported from the Stephen De Hart shelter also (Funkhouser and Webb, 1930: 291), but it was not listed by Goslin in his table of food materials from Kentucky rock shelters, and its age has not been reported.

Corn has been reported from Early Woodland contexts at the Williams site in Gordon County and at 9 PI 103 in Pickens County in northwest Georgia (Morse, 1960: 88), but the suggested antiquity of the corn has not yet been verified. Dan Morse believes that the corn from these two sites and from a site in eastern Tennessee should date between 300 B.C. and 1 A.D. (personal communication, 1962). Thus they would fall within the earliest Middle Woodland time range.

Two Adena mounds in Ohio have produced cultigen remains. Citing Richard G. Morgan's Florence Mound Field Notes of 1938, Goslin (1957: 42) reports that "fragments of pumpkin or squash rind were recovered from the fire place, where they were associated with other food materials and some artifacts." Charcoal from the fireplace on the floor of this mound was radiocarbon dated at A.D. 530 ± 250 (Libby, 1954: 737), but this date seems too late for Adena.

Fragments of pumpkin or gourd and Chenopodium seeds were found on the floor of a burned house beneath the Cowan Creek Mound (UMMAEL Rept. 285). Microscopic examination of the cucurbit rind indicates that it is gourd rather than pumpkin. Charcoal from outside the house in a subfloor fireplace was radiocarbon dated at A.D. 440 ± 250 (Libby, 1952: 78). This also seems to be too late for Adena.

In spite of the paucity of evidence for agriculture from Adena sites, it is commonly thought that the economic system which made the Adena development possible was based on a subsistence pattern which included agriculture. Spaulding (1955: 20) has stated that "It seems equally clear that a reasonably efficient corn agriculture must underlie such an elaborate cultural development. . . . ." If this is true, agriculture should have been practiced by the Adena people near the beginning of their development. The Newt Kash date would seem to support this.

To summarize the postulated early Eastern agricultural complex, it appears to have extended from the Ozarks to Ohio. Gourd, squash, sunflower, *Iva,* perhaps *Chenopodium,* and possibly *Phalaris* were the plants cultivated. Corn and beans are apparently absent since they are from the earlier Tamaulipas agricultural levels in contrast to those of the Southwest. The primary cultural affiliation of the complex appears to be with Adena in Kentucky and Ohio, and it probably originated before 700 B.C. Apparently the only plants of the complex that reached the Upper Great Lakes are squash and sunflower, neither appearing to be very important in most of this region.

The introduction of corn may have been responsible for the decline in importance of sunflower seeds and the eventual eclipse of the other early "grains"(e.g., *Iva* and *Chenopodium*). With the possible exception of a charred nubbin from the Rock Creek Shelter near Gettysburg, Pennsylvania, which may have been associated with potsherds much like Vinette I (UMMAEL Rept. 390), corn occurs earliest in the eastern United States in a few Hopewell sites. Six such occurrences have been reported, but at least two of them are questionable. They stretch east and west from Ohio to Kansas City near the fortieth parallel for about 630 miles but only 90 miles north and south.

In 1882, Putnam and Metz discovered "charred corn and corn-cobs" with four pottery vessels in a fireplace at the center of the base of Mound One of the Turner Group in Hamilton County, Ohio (Willoughby 1922: 29). I have found no record of the disposal of this corn. Four years later in a list of additions to the Peabody Museum, Putnam (1886: 426) mentions corn ". . .from mounds belonging to the Liberty Group on the land of Edwin Harness, esq., in Liberty township, Ross Co., Ohio--Explorations conducted for the Museum by Mr. F. W. Putnam and Dr. C. L. Metz." It has been assumed that this corn is Hopewell, and it probably is, but the exact provenience cannot be established.

It was not until 1937 that Wedel (1938: 101) made the next discovery of Hopewell corn at the Renner site near Kansas City, Missouri. Beans were reported also, marking what is possibly the earliest appearance of this cultigen in the United States outside of the Southwest. "From pit 9, in a layer of charred grass 27 inches beneath the surface, were taken six short, wide kernels of maize. Among other carbonized materials at a depth of 30 inches were identified several small beans resembling modern pintos in size and shape" (Wedel, 1943: 26). Roedl and Howard (1957: 57) report "a kernel of what appears to be maize" from their excavations which are adjacent to those of Wedel. We know of nothing that "appears to be maize" except maize, although many other plant remains may be easily misidentified (including beans).

Radiocarbon analysis has furnished four dates for the Renner Hopewellian occupation. The first date to appear, A.D. 680 $\pm$ 125 (Crane and Griffin, 1958*a*: 119), was rejected by Griffin

(1958: 20) as too young. Three more dates that appeared subsequently are A.D. 440 ± 100, A.D. 110 ± 100, and A.D. 10 ± 100 (Crane and Griffin, 1959: 178-80), averaging A.D. 185.

In 1950 "a few fragments of charred corn" were recovered from a Hopewell village at the Clear Lake site in Tazewell County, Illinois (Fowler, 1952: 169), but a Hopewell provenience for this corn has been questioned. The same year, corn was discovered at the Knight or Ansell site in Calhoun County, Illinois (Plate 1). "Not only was there more than a cupful of charred, well-preserved corn kernels, but mixed with them was a few fragments of charred cornstalks . . . . The charred corn had been thrown into a trash pit in the village site together with several sherds of definite Hopewell type" (McGregor, 1958: 169). This corn was analyzed and described by Cutler as follows:

> Most of the grains are from 12-rowed ears with relatively straight rows . . . . The ears probably resembled the Guatemalan Tropical Flints more than they resembled the historic ears of the region. In all respects the material is similar to that from the upper pre-pottery and lower pottery levels of Tularosa cave, New Mexico (*in* McGregor, 1958:170).

Fortunately one element of the site has been radiocarbon dated. A shell dipper from within mound number eight was found to be 1700 ± 175 years old (Crane, 1956: 667). If the Ansell corn is contemporary with the dated mound, it was grown about A.D. 250 There appears to be a definite possibility, however, that the corn is actually Middle Mississippian (Griffin, 1960*a*: 23).

During the summer of 1963, Olaf Prufer excavated the McGraw Hopewell site in Ross County, Ohio. It was my good fortune to visit this site just as corn was first discovered and to examine the context before its removal. No indication was found that could lead to a conclusion that the corn was not Hopewell.

Ritchie (1944: 126) reports a charred section of corn cob from the Wray component of Point Peninsula in Monroe County, New York. He states that it was in the bottom of an abandoned cache pit where burial 10 was found. The validity of this find appears to be highly questionable, however (Griffin, 1961*a*: 93).

Stuart Struever has recovered seeds from the Stillwell site in southern Illinois which were identified by the writer as *Iva annua* var. *macrocarpa* (Blake) Jackson (marsh elder). In addition, Struever has found a very large quantity of *Chenopodium* seeds at the Snyders site in Calhoun County (1962: 584). It is not yet known, however, that these Hopewell seeds are the result of chenopod cultivation.

The bottle gourd apparently was cultivated by, or at least known to, the Hopewell people. A single carbonized gourd seed was present in a large quantity of plant remains recovered by Struever from the Apple Creek site in Madison County, Illinois. The identification was made by Hugh Cutler. In addition, there exists a pottery vessel in the form of a gourd which was excavated from Brangenberg Mound Number 2 in Calhoun County, Illinois, near the Illinois River (Griffin and Morgan, 1941: 42). The likeness of this vessel to a gourd is unmistakable (Plate 45, p. 173). Griffin estimates the age as about A.D. 200 (personal communication, 1963).

On the basis of the scanty Hopewell cultigen remains described above, a second agricultural complex can be suggested. Its geographical extent can be demonstrated only for a narrow east—west band extending from Kansas City to Columbus, Ohio; but presumably it was present throughout the Hopewellian range, including the Goodall Focus of southern Michigan. Of the assumed cultigens in the early Eastern agricultural complex only marsh elder, gourd, and chenopod (?) can be demonstrated to be present; but we do know that squash and sunflower survived in the East throughout later times. At any rate, corn now appears to be no earlier than Hopewell, and beans may be somewhat later (see below).

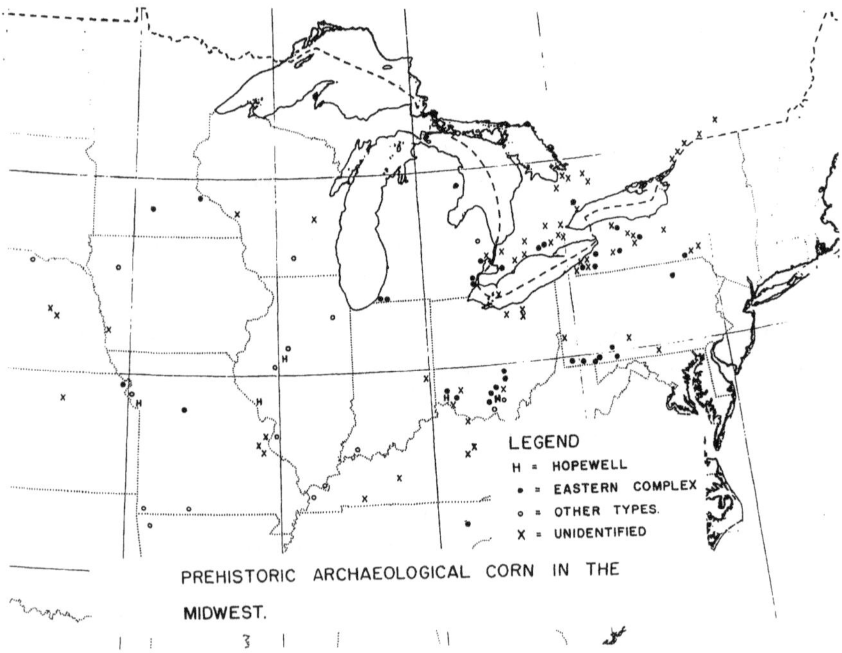

Map 1.

Most of the post-Hopewell archaeological corn remains in the Midwest are, at least to some extent, of a type referred to as Eastern Complex (Carter and Anderson, 1945) or Northern Flint (Brown and Anderson, 1947). We prefer to accept the former name because it sometimes is flour rather than flint and southern rather than northern. It does attain its most distinctive form in the northeastern United States, however, where it is more likely to be flint.

Corn of the Eastern Complex type is perhaps the most easily recognized of any archaeological corn because, in its classic form, both the cobs and the kernels are quite distinctive. Characteristically it has eight rows of kernels with strong row pairing, that is, each row is a little closer to the row adjacent on one side than to the row on the other side. Thus there are four sets of two rows each. This feature usually shows up better on the bare cob than it does on the full ear. There may be ten rows or twelve rows of kernels to the ear, but the row pairing is, nevertheless, usually obvious. The cobs tend to have a distinct square, pentagonal, or hexagonal cross section rather than one which is round or elliptical as in other types. The rows are normally quite straight except in the butt area of the ear which is typically somewhat expanded and attached to an exceptionally large stem or shank. The kernels, which are usually somewhat crescent shaped, may be quite large but are seldom higher than they are wide; and sometimes the height is only half the width. Archaeological kernels usually lack the endosperm area. This accentuates the crescent shape. Kernels are thicker at the apex than near the germ.

The earliest known occurrence of this race is perhaps about 1000 A.D. in Middle Mississippian sites where it is found usually as a minor element in corn which is otherwise much like the "Basketmaker corn" of the Southwest. It appears as the exclusive type of corn only slightly later in Owasco sites in New York.

Eastern Complex corn appears to have its closest relative in the Guatemalan Big Grains. "Varieties with wide kernels, stout cobs, and a more or less enlarged base are practically unknown from most of Mexico. However, they are present in Guatemala and adjacent Chiapas. It seems probable that northern flint corns may be among various cultural traits which spread from south of the Isthmus of Tehuantepec into the eastern United States without leaving any very clear record of the route by which they journeyed" (Brown and Anderson, 1947: 18-19). The classic form of the Eastern type, however, appears to have developed in the eastern United States.

Eastern Complex corn occurs at the Sackett site in New York (Brown and Anderson, 1947: 8, 13) which has a radiocarbon date of A.D. 1140 ± 75 (Crane and Griffin, 1962: 192). This is the earliest dated occurrence east of the Mississippi for beans as well as for classic Eastern corn. There are eight more Owasco sites from which corn has been recovered. Corn of the Eastern type was obtained from Castle Creek and Kahler in New York (Brown and Anderson, 1947: 8, 13), from Lyons Farm in Bradford County, Pennsylvania (UMMAEL Rept. 246), and from Goessens and Krieger in Ontario (UMMAEL Repts. 303 and 289). Corn from the Bainbridge, Bates, and Owasco Lake sites in New York (Ritchie 1944: 68, 83; 1959: 11) has not been identified to race, but presumably it is the same. Beans are reported from Bainbridge and from Castle Creek, the latter site dating A.D. 1200 ± 100 (Crane and Griffin, 1958 *a:* 1100) and A.D. 1430 ± 100 (Crane, 1956: 668).

It is likely that corn and beans reached the Owasco people by way of the Ohio Valley which was occupied by the early Fort Ancient peoples at the time. Baum, Gartner, Baldwin, Feurt, Madisonville, Campbell Island, and Cramer villages produced Eastern corn (Brown and Anderson, 1947: 8, 13; UMMAEL Rept. 209; Anderson *in* Goslin, 1952: 17). Corn from Anderson, Fox Farm, and Turpin Villages (Griffin, 1943: 375; Oehler, 1950: 26) has not been identified to race. Beans were present at Campbell Island, Baum, Gartner, Feurt, Fox Farm, and Cramer; and sunflower seeds were recovered from Cramer and Campbell Island. Squash (or gourd) occurred at Cramer only (Goslin, personal communication, 1963). Turpin Village has been radiocarbon dated at A.D. 1285 ± 75 and Haffner—Kuntz at A.D. 585 ± 75 (Crane and Griffin, 1961: 110-111). The latter date is probably too early (Griffin, personal communication 1962).

Eastern corn was recovered from Canter's and Kettle Hill Caves in Ohio. A radiocarbon date of A.D. 787 ± 100 for Ash Cave, Ohio, was obtained from a "Sample taken from refuse layer ca. 2 to 3 in. below surface in ash bed, 2 to 3 ft. deep. Associated material in the refuse layer included a pocket of ca. 100 pieces of broken pottery, grass, an assortment of gourd rinds, one piece of pumpkin. . .", etc. (Crane and Griffin, 1962: 188). The dated material is Ohio State Museau No. 899/1. Included in their No. 899/2 of Ash Cave are three fragments of eight-row corn cob. Kettle Hill Cave contained pumpkin (squash) and sunflower also, and Canter's Caves produced gourd and squash (UMMAEL Repts. 15, 31, 368).

It is of considerable interest that in Kentucky and Ohio squash and gourd remains have been found almost exclusively in caves.

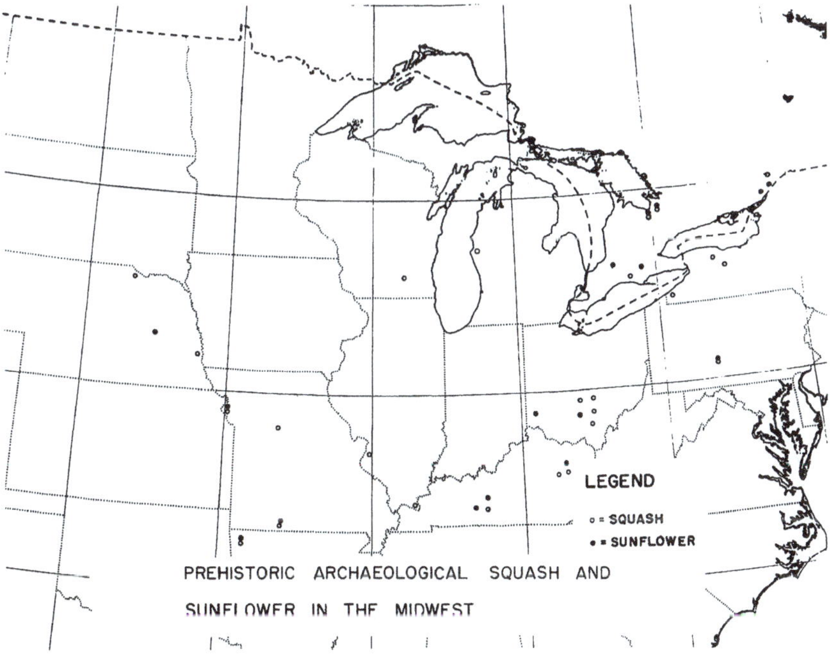

Map 2.

One might consider this to be the result of differential preservation except for the fact that beans were not found in the caves, although they usually occur in sites of the Fort Ancient aspect. Experiments recently carried out by the writer in the UMMA ethnobotanical laboratory strongly indicate that, of the cultigen remains that concern us here, corn, beans, and sunflower seeds are considerably more likely to survive burning and washing than are squash or gourd remains. This is borne out by the fact that, of 11 archaeological finds of gourd east of the Mississippi River, west of the Appalachians, and north of Tennessee, 10 are unburned materials from caves. Of 9 sites yielding sunflower seeds in this area, 3 are cave sites, and 5 are protohistoric sites in Ontario. Of 18 sites where squash remains were found, 8 are in caves, 5 are protohistoric, and 3 are Middle Mississippian. All of the 18 archaeological sites from which beans were recovered are in the open, as are the great majority of the sites producing corn.

Differential preservation would be a more acceptable explanation of the mutually exclusive distributions of beans and squash-gourd if it could be assumed that beans were not grown in the East before the appearance of the Fort Ancient culture and that the main occupations of the caves took place in earlier times. There is a report, however, of Hopewell beans from the Renner site. Unfortunately they could not be preserved for positive identification. Charred seeds of another legume, the honey locust, are almost identical in outward appearance to charred beans except that the hilum is on the end rather than on the side of the seed. It is easy to confuse the identification of these two seeds, especially when they are not well preserved. Many remains of honey locust pods were recovered from Newt Kash Hollow and the Ozark Bluff shelters (Jones, 1936 *b*: 154). Charred honey locust seeds were found at Cahokia and in the Bynum Mound (UMMAEL Repts. 302, 261). Those from the latter site were thought to be beans until Volney Jones correctly identified them.

Apparently, there is only one pre-Iroquoian site in the northeastern United States in which both beans and squash, or gourd, occur. Beans and squash and also corn were recovered from the Lynch site in Boyd County, Nebraska (UMMAEL Rept. 94). This site has been radiocarbon dated at A.D. 1140 ± 100 and A.D. 1420 ± 100 (Crane and Griffin, 1962: 194).

If beans did not enter the east until after 1000 A.D. (perhaps from the Southwest), it would explain their absence from the Sterns Creek level of the Walker-Gilmore site in Cass County, Nebraska, where squash *(Cucurbita pepo)* and gourd remains were plentiful (Strong, 1935: 193) at A.D. 930 ± 150 (Crane and Griffin, 1962: 193). Explaining the absence of corn is another problem. It may be that corn was not grown there for a combination of environmental and technological reasons. Possibly corn culture simply had not yet reached that far up the Missouri River (it had however, 800 years or so to reach there from the Kansas City area.) Kernels of a "relatively primitive" type of corn have been found in Platte County, Nebraska, in the Lawson Component of the Loseke Creek Focus, but it apparently is younger than Sterns Creek (Kivett, 1952: 58, 68; Wedel, 1959: 550).

In southwestern Pennsylvania generalized Eastern Complex kernels were found in an Early Monongahela context at the Varner site (UMMAEL Rept. 412). The approximate date of the cultural period is estimated at A.D. 900 to 1100 by Mayer-Oakes (personal communication, 1962). I found similar kernels among materials recently recovered from the Juntunen site at the western tip of Bois Blanc Island in Lake Huron, a few miles east of the Straits of Mackinac. These kernels are mostly wider than high but are rather thick tipped. Thus they may be an early form of Eastern corn (Fig. 1).

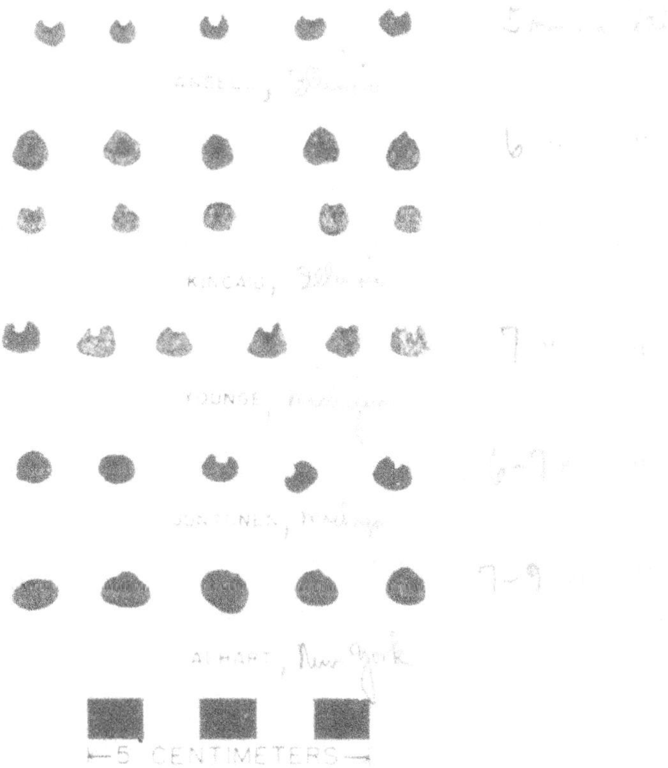

Fig. 1. Corn from five Archaeological sites in the Midwest.

The radiocarbon dates for this site are A.D. 835 ± 75, A.D. 910 ± 75, A.D. 1070 ± 75, and A.D. 1320 ± 75 (Crane and Griffin, 1961: 110; 1963: 233). Most of the corn appears more likely to be contemporary with the A.D. 1330 date than with the others, because it generally occurs in the upper levels of the site; but some of it may be as early as A.D. 1070.

Besides Eastern Complex, there is another race of corn represented from Canter's Caves in Ohio which appears to be derived from the Basketmaker (Pima–Papago) race. The cobs are elliptical in cross section with small shanks and taper toward both ends. There is no row-pairing in this race, and row number is usually 12 or 14. Griffin (1949: 83) has mentioned the Basketmaker corn found "in the Ozark Bluff Shelters at an unknown time and in Canters Cave, Ohio, in a Woodland complex roughly of the Hopewell period. This could be the earliest corn in the east." In regard to the contemporaneity of Hopewell and Canter's

Cave corn, he points out that "this association is not definite since Fort Ancient pottery was also taken from the site. All Fort Ancient corn is, however, of the Eastern type."

Perhaps the oldest dated Basketmaker corn in the East is that found near St. Louis at Cahokia, along with squash. Most of it came from a pit beneath the slope of Mound 34. This is the same pit that produced the material which was radiocarbon dated at A.D. 1150 ± 150 (Crane, 1956: 666). A date of A.D. 1290 ± 100 postdates Mound 34, and a date of A.D. 1300 ± 100 should equate with Mound 34 in time. Two dates on materials from 150 yards east of Monks Mound are A.D. 1000 ± 125 and A.D. 1480 ± 100 (Crane and Griffin, 1959: 181-182). Part of the Cahokia corn has Eastern characteristics, that is, 12 cob fragments out of 94 are largely Eastern Complex in nature. But the rest of them exhibit a stronger Basketmaker element. Mean row number for the 94 cob segments is 11.1 (10 with 8 rows, 31 with 10 rows, 46 with 12 rows, and 7 with 14 rows). The cob diameters, exclusive of the glumes, average 12.7 mm. Most of the kernels are not Eastern but tend to be rather small, irregular, and somewhat globular (UMMAEL Rept. 302).

Cobs at the U. S. National Museum which were recovered from the Steed-Kisker site near Kansas City have 8 to 12 rows with no row-pairing. They taper toward each end, and are not like the northeastern flints (Brown and Anderson, 1947: 9, 13). Steed-Kisker kernels at the UMMA Ethnobotanical Laboratory are perhaps more Eastern in appearance than those from Cahokia as are some of the cob fragments, but some of them are far from typical Eastern Complex. Two of 6 cobs have 8 rows, and 4 cobs have 10. Squash and sunflower seeds occurred at Steed-Kisker also (UMMAEL Rept. 115). A recent date from the University of Michigan Radiocarbon Laboratory for the Warren Gresham site of the Steed-Kisker Focus is A.D. 887 ± 150 (Sample M-1182).

The Crable site of Fulton County, Illinois, has yielded cobs which are similar to, but not typical of, Eastern corn; cob diameter averages 12.5 mm, and row number averages 11.0 (nine cobs with 10 rows, six with 12, and one with 14). This is almost the same as the figures for the Cahokia cobs and would not allow enough space for the broad kernels of classic Eastern corn (UMMAEL Rept. 61). Three radiocarbon dates from Crable charcoal are A.D. 1340 ± 100, A.D. 1360 ± 100, and A.D. 1430 ± 100. An A.D. 800 ± 100 date from mussel shell probably is too early. "The Crable Site is a large Spoon River Focus, Middle Mississippi Phase village. . . in the Central Illinois valley there is clear evidence of an Early Mississippi stage derived from the Old Village level at Cahokia, and this occupation has not yet been dated. The Crable dates indicate that by A.D. 1300 the Spoon River Focus had assumed its distinctive form" (Crane and Griffin, 1959: 181).

Still more nearly Eastern is the corn from the Kingston Village of Peoria County, Illinois, and to some extent, the Kincaid site of Pope and Massac Counties, Illinois. The few Kingston cob fragments, while not typical of the classic Eastern cobs, are similar in some respects, one having 10 rows and another 12. Many of the Kingston kernels are distinctly Eastern, but some are much more like those from Cahokia. Others are rather large and wedge-shaped like the Kincaid kernels. There are also kernels that look quite like those from the Fisher site. The Kincaid kernels, some of which are comparable to Eastern kernels in size, are more wedge-shaped than crescentic and are not as low and rounded (Plate 1). Some of them look like Eastern kernels that did not have space enough to expand laterally, but generally they are quite southwestern in appearance. The Kincaid cobs are larger than those from Cahokia and Crable and well within the Eastern range, but row number averages rather high (9 cobs with 10 rows, 3 with 12, and 1 with 14, averaging 10.8 or about the same as Cahokia and Crable). Some of these cobs are quite Eastern in appearance, but the majority are not. Squash or gourd was found at Kincaid (UMMAEL Repts. 411 and 67).

At the Angel site in southwestern Indiana, kernels were recovered that are generally Eastern but are not so broad as those of classic Eastern complex corn. Most of these kernels are about as wide as they are high and are more or less circular in outline. The Angel site also contained beans (UMMAEL Rept. 293).

Archaeological corn has been recovered from several Middle Mississippian and Caddoan sites to the south of Cahokia. Four of these are Wickliffe in Ballard County, Kentucky; Banks in Crittendon County, Arkansas; Belcher in Caddo County, Louisiana; and Davis in Cherokee County, Texas. Brown and Anderson (1947: 8, 13) indicate that the Wickliffe corn is not Eastern Complex, but they mention one 8-row cob and two cobs each with 12 and 14 rows of kernels. Kernel width ranges from 7 to 10 mm but is mostly on the narrow side. Thus it is likely that this corn is similar to that of Cahokia in being mostly Basketmaker-like but with some Eastern elements. This is certainly true for the corn from Banks, Belcher, and Davis which I have examined. The Davis corn has been radiocarbon dated at A.D. 400 ± 175 (Libby, 1952: 83), and because of this early date has been considered extremely significant in discussions of the history of aboriginal corn cultivation in the eastern United States. However, a recent date of A.D. 1307 ± 75 from the University of Michigan Radiocarbon Laboratory (M-1186) indicates that the Davis corn is no more significant than the corn from most other Middle Mississippian sites. The corn itself, in my opinion, is of essentially the same type as other Middle Mississippian corn (Griffin and Yarnell, 1963).

Corn has not been reported from Aztalan, but remains of squash have been found there (Barrett, 1933: 115, 128, 356). There are 2 radiocarbon dates for Aztalan, A.D. 1625 ± 100 and A.D. 760 ± 75 (Crane and Griffin, 1962: 187). The former date is too late, but the latter may be too early. The Dietz site near Madison, Wisconsin, appears to be ceramically related to the Aztalan development and also to the Effigy Mound culture. Its date has been estimated at about A.D. 1250 (Baerreis and Nero, 1956: 17). Cutler (1956: 17) reports that the Dietz corn "is more like the older kinds of corn grown in the Upper Mississippi Valley than the late Indian kinds, and is similar to even older materials in Mexico, New Mexico, Arizona and Illinois It is like some Hopewell corn from Site 17, Ansell and Knight farms, Calhoun Co., Illinois, in row number and grain thickness. Some modern tropical flints of Central and South America are not greatly different." The Ansell corn, however, may be Middle Mississippian rather than Hopewell.

It appears that corn of the Basketmaker race entered the East in the southern Illinois area before the Mississippian expansion (Griffin, 1946: 60), presumably in Hopewell times. Its principal occurrence is at Cahokia where its characteristics are much more obvious than those of the Eastern Complex race. The greater the distance south, north and east from Cahokia the weaker is its influence until we reach Ohio where the Fort Ancier sites exhibit corn that is definitely Eastern but with some kernels and cobs much like those from Kincaid with non-Eastern tendencies. The occurrence of Basketmaker cobs in Canter's Caves indicates that the race made an early appearance in Ohio, and suggests that what we are calling Basketmaker is actually an early type of corn in the East which survived later in the vicinity of the Mississippi and Missouri rivers where the climate is drier Assuming that the Basketmaker race is adapted to rather warm and dry climates, any westward recession of such a climate migh cause a westward recession of that race in favor of a corn better adapted to cooler and wetter conditions. If Eastern Complex evolved from a type that originated in highland Guatemala, it may have been already adapted to recent climatic conditions of the Northeast.

Galinat and Gunnerson (1963) suggest that Eastern Complex corn originated in a blend of the Chapalote and Harinoso de Ocho races with teosinte in the southwest. "As this maize spread northward into the Dakotas in the Plains east of the Rocky Mountains and east across northern United States and on into New England, there was a filtering out of the eight-rowed element which became stabilized as the race called Northern Flint "(p. 120). According to these authors, Harinoso de Ocho apparently arrived in the southwest from northwest Mexico at about A.D. 700, after which time eight-row corn spread to the East.

Upper Mississippian corn appears to be essentially Eastern Complex. Kernels from the Fisher Oneota site in northeastern Illinois are Eastern in shape but smaller than typical kernels of that race, 40 kernels averaging 7.2 mm in width (UMMAEL Rept. 75). Two caches of Eastern corn were found at the Moccasin Bluff site in southwestern Michigan where some Oneota potsherds were found (Hale G. Smith, Moccasin Bluff Field Notes, UMMA). McKern (1945: 138) has reported two occurrences of Oneota corn from Wisconsin for which there is no morphological description. These are from the McCauley site (Lake Winnebago Focus) in Oshkosh and Shrake-Gillies site (Orr Focus) of Trempealeau County. Corn has also been reported from Carcajou Point in Jefferson County (Hall, 1962: 31).

Two Upper Mississippian sites in southern Minnesota have produced classic Eastern complex corn. These are Silvernale (Cannon Junction) near Red Wing and Cambria (Jones Village) near Cambria (UMMAEL Rept. 292).[2] The same type of corn is found in southern Oneota sites, notably at Utz in Missouri and Fanning in Kansas. The Utz site produced squash and beans in addition to corn (UMMAEL Repts. 293 and 372). Sites in Missouri, Kansas, Nebraska, the Dakotas, Iowa, and Illinois where the Eastern race occurs exclusively are protohistoric and historic. The kernels in these cases are often thicker and wider than those from sites farther east.

In southwestern Pennsylvania the corn from five Late Monongahela sites is Eastern Complex. These sites are Emerick (which also yielded beans), Reckner, Powell 1, Martin, and Phillips (UMMAEL Rept. 229). Corn, beans, squash, gourd, and sunflower

[2] Although Cannon Junction has an "Old Village" component (Griffin, 1960: 838) the corn from this site is strongly Eastern Complex like corn from late sites throughout the greater Midwest.

seeds have been recovered from Sheep Rock Shelter in Huntingdon County (Kinsey, 1960: 17; Cutler and Whitaker, 1961: 472).

Eastern corn has been recovered from a cache near Mayville and from a hilltop fort near Ellington in Chautauqua County, New York (UMMA collections). Cyrus Thomas (1894: 511, 514, 515) reports corn from a pit within a circular earthwork near Fredonia in Chautauqua County, from a pit within a hilltop fort near Portage in Wyoming County, and in pots with burials near Lima in Livingston County, New York. Other sites in Chautauqua County from which cultigen remains have been recovered are Ripley with corn, beans, and squash (Parker, 1907: 547), McCullough with corn (Parker, 1908: 97), and Westfield with corn and beans (Guthe, 1958: 39). The Richmond Mills site of Ontario County yielded corn, beans, and squash (Parker, 1918: 16), and the Wilkins site of Orleans County produced corn (Vanderlaan, 1960: 5). The Silverheels corn of Erie County is Eastern Complex (Brown and Anderson, 1947: 8, 13), and so is the Alhart corn of Monroe County (Plate 1) which was accompanied by squash and beans (UMMAEL Rept. 2).

Some of the charred Alhart kernels appear to be dented. Although there are historical records of dent corn in Virginia and New Jersey, there are no reports of archaeological dent corns from east of the Mississippi (Brown and Anderson, 1948: 256). Our experiments have shown that heated kernels usually lose their dents and that dry and wrinkled sweet corn becomes smooth when heat is applied. Apparently there are no archaeological kernels from east of the Mississippi that have not been charred.

Corn has been recovered from many Iroquoian sites in Ontario; but the type is known in only four cases, all of which are Eastern Complex. These are the Seed site, a prehistoric village in Vaughn Township, York County, a site near Aylmer, Elgin County, which also produced squash and beans, Dutton's Ash Pit, an historic Huron site with beans (UMMAEL Repts. 288, 176, 313), and the Miller site, Ontario County (W. A. Kenyon, personal communication, 1962; identified by Yarnell).

Several other southern Ontario sites, all with corn, contained the remains of other cultigens. Squash, beans, and sunflower seeds were found in the Roebuck site in Grenville County (Wintemberg, 1936: 12-13). Remains of these cultigens were found also in Simcoe County at the Flanagan site (Jury, 1948: 6) and at Sidey-Mackay Village (Wintemberg, 1946: 155). Sunflower seeds were recovered from the Uren site in Oxford County (Wintemberg, 1928: 4). This site represents "the earliest level of Iroquoian development in southern Ontario" (Griffin, 1952: 188). Beans and sunflower seeds were found at the Lawson site in Middlesex County (Wintemberg, 1939: 8), and the related Clearville site in Kent County produced beans (Jury, 1941: 3, 5). Three quarts of beans and corn were taken from a mound on Pelee Island in western Lake Erie (Boyle, 1899: 12). In addition, there is corn from about twenty sites in Glengary, Dundas, Grenville, Victoria, Simcoe, York, Waterloo, Brant, Oxford, Middlesex, and Kent Counties, not counting six postcontact villages in Simcoe County. There appears to be no report of archaeological gourd from Ontario nor from anywhere nearer the Great Lakes than southern Ohio.

Four Whittlesey Focus sites have produced corn; none of this has been identified to race. They are the Tuttle Hill and South Park sites in Cuyahoga County, Ohio (Greenman, 1937 *a*: 311), the Franks site in Lorain County, Ohio (Vietzen, 1945), and the Wolf site in Macomb County, Michigan (Greenman, 1939: 7). In the same general region around the west end of Lake Erie, a pit near the Gibraltar site and a cache in the Ford Turf Garden, both in Wayne County, Michigan, produced classic Eastern Complex corn (UMMAEL Rept. 324; UMMA Collections). The Gilbraltar corn was radiocarbon dated at A.D. 1600 ± 200 (Crane, 1956: 668). Beans were found in site 20 MB 33 in Macomb County by Arnold Pilling (UMMAEL Rept. 403). The Parker Earthwork enclosure, two miles east of the St. Clair River in Ontario, produced corn (Lee, 1958: 13). A site (20 MR 5) near Stoney Creek in Monroe County, and the Verchave II site (20 MB 181) in Macomb County, Michigan, each contained a few kernels (UMMA Collections).

The only evidence of possible influence from Middle Mississippian corn in Michigan is from the Younge site in Lapeer County. The kernels include several quite like those from Kincaid (Plate 1). It is similar to Eastern Complex corn but much more "Mississippian" than corn from other Michigan sites. Thus it may be the earliest archaeological corn from Michigan (with the possible exception of the Juntunen corn). The Younge site appears to have strong cultural ties with the Lawson site of Ontario (Kidd, 1952: 74). Wintemberg (1939: 8) reports that some of the Lawson cobs are oval in cross section, like Basketmaker corn.

Several charred cobs of Eastern corn were found by Fred Dustin three to four feet deep on a firebed of stones at Green Point at the junction of the Tittabawasee and Shiawasee Rivers near Saginaw, Michigan (UMMA Collections). One unburned Eastern cob was taken from Great Warren Dune in Berrien County where it was found with Woodland potsherds, large amounts of mammal bone, and a hearth (UMMA, Birdsell Collection).

At the Dumaw Creek Site in Oceana County near Lake Michigan, Quimby (personal communication, 1961) found squash seeds in a net bag with a very late Woodland burial but with no evidence of European contact. The radiocarbon date on material which included two of the unburned seeds is A.D. 1680 ± 75 (Crane and Griffin, 1961: 110).

In material submitted for radiocarbon dating from the Mikado (Pine River) enclosure in Alcona County, Michigan, is a small, immature, 8-rowed Eastern cob. This seems to be the only corn associated with a Michigan earthwork. The charcoal sample from the base of the bank of this enclosure was dated A.D. 1410 ± 100 (University of Michigan Radiocarbon Laboratory records). Two other circular enclosures in the northern Lower Peninsula have been radiocarbon dated. The Linseman-Walters enclosure in Ogemaw County was dated A.D. 1360 ± 75 (Crane and Griffin, 1960: 33). A small burial mound near the enclosures at the Missaukee Preserve in Missaukee County was dated A.D. 1210 ± 75 (Crane and Griffin, 1961: 109). Apparently these earthworks are the remains of palisaded villages of the general northeastern type. The Mikado corn appears to be later than the corn from Bois Blanc Island.

In summary it can be said that gourd, squash, sunflower, and marshelder appear to be earlier than corn and beans in the eastern United States. Corn would seem to have made its first appearance before A.D. 100, but beans do not appear east of the Mississippi until after A.D. 1000.

Sunflower seeds have been found in the Ozark Bluff shelters, Newt Kash Hollow, and Mammoth Cave. In the prairie region they occurred at Steed Kisker and elsewhere in the Missouri Valley and at the Larsen site near the lower Loup and Platte Rivers. In the northeast they have been recovered from 2 Fort Ancient sites, from Sheep Rock Shelter in Pennsylvania, and from 6 Iroquoian sites in Ontario (but not from Owasco sites). Domesticated sunflower seeds are rare or nonexistent in archaeological sites elsewhere. For this and other reasons, it appears likely that the sunflower was domesticated somewhere in the general Kentucky— Arkansas region, perhaps in the Mississippi Valley.

Gourd remains are from caves and rock shelters with two or three exceptions: Cowan Creek Mound, Sterns Creek, and perhaps Kincaid. They have not been found in the Lakes region, but there is a Hopewell clay effigy gourd from the Illinois Valley. Squash and pumpkin remains are from cave sites, the Florence

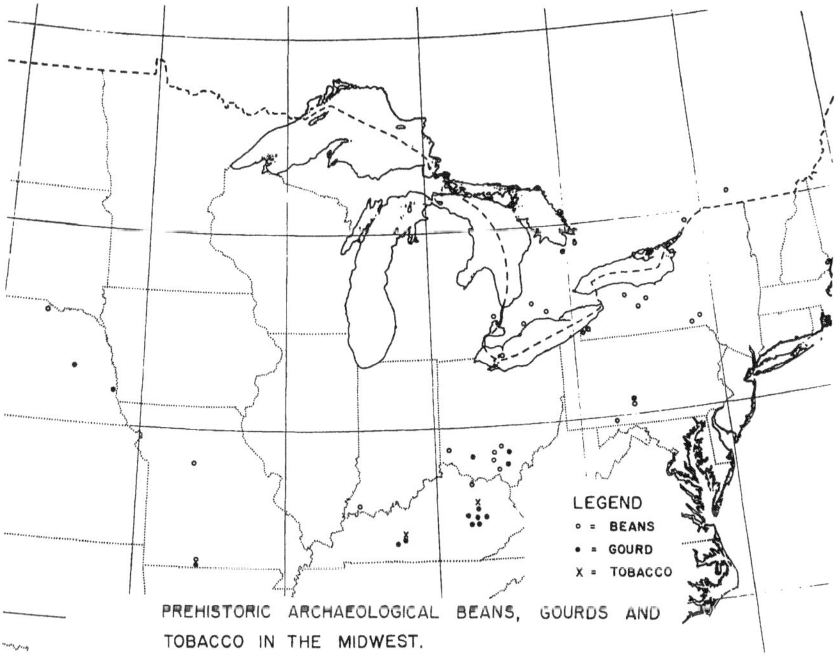

Map 3.

Mound, Sterns Creek, Lynch, 5 or 6 Mississippian sites, 7 Iroquoian sites, and the Dumaw Creek site.

Beans have been found at only 1 Middle Mississippi site, but they have been recovered from 6 Fort Ancient sites, 3 Owasco sites, and 14 Iroquoian sites. In fact, 25 of our 30 Eastern records are from Ohio and northeastward. Beans may have arrived from the Southwest about 1000 A.D.

The earliest corn in the East seems not to have been Eastern Complex, but derived from an early Southwestern Basketmaker type. Eastern corn may have come from the Guatemalan highlands and may have been brought by Mayan speakers up the east coast of Mexico from which it diffused northward through Texas to the Mississippi Valley and up the Ohio Valley to New York and the Upper Great Lakes region. It seems more likely, however, to have arrived from the Southwest along with beans or perhaps earlier. The Middle Mississippi peoples grew both Eastern and Basketmaker corn to varying degrees. Later peoples and more northeasterly peoples apparently grew only Eastern Complex corn.

Noncharred cultigen remains occur only at late sites and cave sites. Thus the only early material from open sites is that which happened to be burned. Gourd and squash usually burn readily to ash, but beans, corn, and sunflower seeds often are preserved by carbonization and thus leave a better record. It is likely that gourd and squash were more widespread than the evidence indicates, but that beans and corn are relatively well represented.

In the Upper Great Lakes region there is archaeological evidence of corn and squash in Ontario, Michigan, and Wisconsin; of beans in Ontario and immediately adjacent Michigan; and of cultivated sunflower in Ontario. Aboriginal cultigens probably arrived in Wisconsin from Illinois in Mississippian times, though corn may have been there in Hopewell times, temporarily if not permanently. Michigan and Ontario corn is quite similar to the corn of Ohio and New York.

TABLE XIII

ARCHAEOLOGICAL REMAINS OF CULTIVATED PLANTS IN AND AROUND THE MIDWEST

Site	Culture	State	County	Cultigen Remains	Date
Harlan Rock Shelter		Ky.	Harlan	C	
Newt Kash Hollow	(Adena?)	Ky.	Menifee	Z? N C L H	700 $\pm$ 150 B.C.
Salts Cave		Ky.	Edmonson	Z N? C L H	1125 $\pm$ 140 B.C.
Stephen DeHart Cave	(Adena?)	Ky.	Powell	Z C L	
Mammoth Cave		Ky.	Edmonson	L H	420 $\pm$ 60 B.C.
					280 $\pm$ 40 B.C.
Little Ash Cave	(Adena?)	Ky.	Lee	L	
Dillard Stamper Cave	(Adena?)	Ky.	Wolfe	L	
Worth Creech Cave	(Adena?)	Ky.	Wolfe	L	
Red Eye Hollow	(Adena?)	Ky.	Lee	L	
Canter's Caves		Ohio	Jackson	Eb C L	
Ash Cave		Ohio	Hocking	Z C L	790 $\pm$ 200 A.D.
Kettle Hill Cave		Ohio	Fairfield	E C H	
Florence Mound	Adena	Ohio	Pickaway	C	(530 $\pm$ 250? A.D.
Cowan Creek Mound	Adena	Ohio	Clinton	L	(440 $\pm$ 250? A.D.
Renner	Hopewell	Mo.	Platte	Z P?	10, 110, 440 $\pm$ 100 A.D.
Ansell	(Hopewell?)	Ill.	Calhoun		250 $\pm$ 150 A.D.
Brangenberg	Hopewell	Ill.	Calhoun	L (clay effigy)	200 A.D.
McGraw	Hopewell	Ohio	Ross	Z	
Clear Lake	(Hopewell?)	Ill.	Tazewell	Z	
Apple Creek	Late Hopewell	Ill.	Madison	L	
Turner	Hopewell	Ohio	Hamilton	Z	
Harness	Hopewell	Ohio	Ross	Z	
Wray	(Pt. Peninsula?)	N.Y.	Monroe	Z	
Rock Creek	(Vinette I?)	Pa.	Adams	Z	
Walker-Gilmore	Sterns Creek	Neb.	Cass	C L	930 $\pm$ 75 A.D.

TABLE XIII *(CONT.)*

Site	Culture	State	County	Cultigen Remains	Date
Lawson	Loseke Creek	Neb.	Platte	*b*	
Varner	Early Monongahela	Pa.	Greene	E	(1000 A.D.?)
Steed-Kisker	Middle Mississippi	Mo.	Platte	*e*B C H	(887 ± 75 A.D.)
Cahokia	Middle Mississippi	Ill.	St. Clair	*e*B C	1150 ± 150 A.D.
Harrell	Middle Mississippi	Mo.	Jefferson	Z	
Long	Middle Mississippi	Mo.	Jefferson	Z	
Boyce	Middle Mississippi	Mo.	Jefferson	Z	
Aztalan	Middle Mississippi	Wis.	Jefferson	C	1630 ± 100 A.D. 760 ± 75 A.D.
Dietz	Middle Mississippi	Wis.	Dane	b	(1250 A.D.?)
Davis	Gibson Aspect	Texas	Cherokee	*e*b H	400 ± 175 A.D. 1307 ± 75 A.D.
Phipps	Mill Creek	Iowa	Cherokee	*e*b	1100 ± 75 A.D.
Angel	Middle Mississippi	Ind.	Vanderburgh	*e*b P	
Kingston	Middle Mississippi	Ill.	Peoria	*e*b	
Crable	Middle Mississippi	Ill.	Fulton	*e*b	1340, 1360, 1430 ± 100 A.D.
Kincaid	Middle Mississippi	Ill.	Pope and Massac	*e*b C or L	1285 ± 75 A.D.
Wickliffe	Middle Mississippi	Ky.	Ballard	*e*b	
Fisher		Ill.	Will	Eb	
Carcajou Point	Oneota	Wis.	Jefferson	Z	1000, 1030, 1530 ± 125 A.D.
McCauley	Oneota	Wis.	Winnebago	Z	
Shrake-Gillies	Oneota	Wis.	Trampealeau	Z	
Fanning	Oneota	Kans.	Doniphan	E	
Utz	Oneota	Mo.	Saline	E P C	
Cambria	Cambria	Minn.	Blue Earth	E	

TABLE XIII *(CONT.)*

Site	Culture	State	County	Cultigen Remains	Date
Silvernale	Middle Mississippi and Oneota	Minn.	Goodhue	E	
Lynch	Oneota	Neb.	Boyd	E *b* P C	1140, 1420 ± 100 A.D.
Kullbom		Iowa	Mills	*e*	
Larsen		Neb.	Platte	Z P H	
"Ratcliffe"		Ind.	Wayne	Z	
Baum	Fort Ancient	Ohio	Ross	E P	
Gartner	Fort Ancient	Ohio	Ross	E P	
Cramer	Fort Ancient	Ohio	Ross	E P C or L H	
Baldwin	Fort Ancient	Ohio	Fairfield	E	
Feurt	Fort Ancient	Ohio	Scioto	E *b* P	
Anderson	Fort Ancient	Ohio	Warren	Z	
Campbell Island	Fort Ancient	Ohio	Butler	E P H	
Madisonville	Fort Ancient	Ohio	Hamilton	E	
Turpin	Fort Ancient	Ohio	Hamilton	Z	1285 ± 75 A.D.
Fox Farm	Fort Ancient	Ky.	Mason	Z P	
Junturen	Straits Focus of Late Woodland	Mich.	Mackinac	E	1070-1320 ± 75 A.D.
Younge	Owasco	Mich.	Lapeer	*eb*	
Mikado		Mich.	Alcona	*e*	1410 ± 100 A.D.
Martin	Late Monongahela	Pa.	Fayette	E	
Phillips	Late Monongahela	Pa.	Fayette	E	
Reckner	Late Monongahela	Pa.	Somerset	E	
Powell 1	Late Monongahela	Pa.	Somerset	E	
Emerick	Late Monongahela	Pa.	Somerset	E P	
Buchanan Shelter	Late Monongahela	Pa.	Beaver	Z	
Sheep Rock Shelter		Pa.	Huntingdon	Z P C H	
Lyons Farm	Owasco	Pa.	Bradford	E	

TABLE XIII *(CONT.)*

Site	Culture	State	County	Cultigen Remains	Date
Castle Creek	Owasco	N.Y.	Broome	E P	1200, 1430 ± 100 A.D.
Bates	Owasco	N.Y.	Chenango	Z	
Bainbridge	Owasco	N.Y.	Chenango	Z P	
Owasco Lake	Owasco	N.Y.	Cayuga	Z	
Sackett	Owasco	N.Y.	Ontario	E P	1140 ± 75 A.D.
Kahler	Owasco	N.Y.	Livingston	E	
Goessens	Owasco	Ont.	Oxford	E	
Krieger	Owasco	Ont.	Kent	E	
Miller		Ont.	Ontario	E	
Stafford		Ont.	Elgin	Z	(pre-Uren)
"Mayville"	Iroquoian?	N.Y.	Chautauqua	E	
"Ellington"	Iroquoian?	N.Y.	Chautauqua	E	
"Fredonia"	Iroquoian?	N.Y.	Chautauqua	Z	
McCullough	Iroquoian	N.Y.	Chautauqua	Z	
Westfield	Iroquoian	N.Y.	Chautauqua	Z P	
Ripley	Iroquoian	N.Y.	Chautauqua	Z P C	
Silverheels	Iroquoian	N.Y.	Erie	E	
"Portage"	Iroquoian	N.Y.	Wyoming	Z	
"Lima"	Iroquoian	N.Y.	Livingston	Z	
Richmond Mills	Iroquoian	N.Y.	Ontario	Z P C	
Alhart	Iroquoian	N.Y.	Monroe	E P C	
Wilkins	Iroquoian	N.Y.	Orleans	Z	(1300 A.D.)
Hochelaga	Iroquoian	Quebec	Laval	Z P	
"Markley"	Iroquoian	Ont.	Dundas	A	
"McLennan"	Iroquoian	Ont.	Glengarry	Z	
Roebuck	Iroquoian	Ont.	Grenville	Z P C H	
"Reade"	Iroquoian	Ont.	Grenville	Z H	
Dutton's Ash Pit	Iroquoian	Ont.		E P	

TABLE XIII (CONT.)

Site	Culture	State	County	Cultigen Remains	Date
"Corsons"	Iroquoian	Ont.	Victoria	Z	
"Fenelon" 23	Iroquoian	Ont.	Victoria	Z	
Flanagan	Iroquoian	Ont.	Simcoe	Z P C H	
Sidey-Mackay	Iroquoian	Ont.	Simcoe	Z P C H	
Seed	Iroquoian	Ont.	York	E	
Lawson	Iroquoian	Ont.	Middlesex	Z P H	
Coleman	Iroquoian	Ont.	Waterloo	Z	
Middleport	Iroquoian	Ont.	Brant	Z	
Uren	Iroquoian	Ont.	Oxford	Z H	
"Aylmer"	Iroquoian	Ont.	Elgin	E P C	
Clearville	Iroquoian?	Ont.	Kent	Z P	
Pelee Island		Ont.	Essex	Z P	
Parker	Whittlesey?	Ont.	Lambton	Z	
Tuttle Hill	Whittlesey	Ohio	Cuyahoga	Z	
South Park	Whittlesey	Ohio	Cuyahoga	Z	
Franks	Whittlesey	Ohio	Lorain	Z	
Wolf	Whittlesey	Mich.	Macomb	Z	
20 MB 33		Mich.	Macomb	P	
Ford Turf Garden		Mich.	Wayne	E	
Gibraltar	Late Woodland	Mich.	Wayne	E	1640 $\pm$ 100 A.D.
Green Point		Mich.	Saginaw	E	
Moccasin Bluff	Late Woodland and Fisher Focus	Mich.	Berrien	E	
Great Warren Dune		Mich.	Berrien	E	
Dumaw Creek	Late Woodland	Mich.	Oceana	C	1680 $\pm$ 75 A.D.
20 MR 5	Late Woodland	Mich.	Monroe	Z	
Verchave II (20 MB 181)	Late Woodland	Mich.	Macomb	E	

Note: Z - Corn *(Zea mays)*; E - Eastern Complex corn; B - "Basketmaker" corn (non-Eastern Complex); Lower case *e* and *b* indicate minor or uncertain affiliation; N - Tobacco *(Nicotiana)*; P - Beans *(Phaseolus vulgaris)*; C - Squash; Pumpkin *(Cucurbita pepo)*; L - Gourd *(Lagenaria siceraria)*; H - Sunflower *(Helianthus annuus)*.

VI

CLIMATE AND THE DISTRIBUTION OF PREHISTORIC AGRICULTURE IN THE MIDWEST

This study of climatic factors in relation to distribution of agricultural sites is directed primarily toward ascertaining limiting factors along the northern edge of the region of aboriginal agriculture in the Eastern United States and in the Great Lakes region in particular. It considers, however, a large area to the south of this region for comparative purposes.

The area studied lies between central Minnesota and Montreal at the north and between the lower Missouri Valley and the Appalachian Mountains toward the south. It extends from southern Kentucky at the latitude of the confluence of the Ohio and Mississippi Rivers to the Straits of Mackinac.

This region contains more than 120 excavated archaeological sites from which the remains of corn, beans, squash, gourd, and sunflower have been recovered. Of 110 prehistoric sites considered in this study, corn was taken from at least 100, beans from 26, squash from 23, gourd from 13, and sunflower from 12. The cultural assemblages of 38 sites are Iroquoian, including 17 in Ontario, 12 in New York, 5 in Michigan, and 3 in Ohio. Of 9 Owasco sites, 7 are in New York, 1 is in Ontario, and 1 is in Pennsylvania. Also in New York there is 1 Point Peninsula site that yielded corn, according to Ritchie (1944: 126), though it probably is from a later occupation.

In Pennsylvania there are 6 Monongahela sites from which cultigens have been recovered. There are 9 sites of the Fort Ancient Aspect in Ohio and 1 in Kentucky. Also in Ohio there are 2 Adena sites, and in Kentucky there are 6 cave sites said to be or implicated as Adena but are questionable as such. There are 3 additional cave sites in Kentucky, 3 in Ohio, and 2 in Pennsylvania. Corn is reported from 6 Hopewell sites, 2 in Ohio, 3 in Illinois, and 1 in Missouri. In 3 of these sites, however, the corn may be from later occupations. Of 10 Middle Mississippian sites 4 are in Illinois, 2 each are in Missouri and Wisconsin, and 1 each is in Indiana and Kentucky. There are 8 Upper Mississippian sites, including 2 in Minnesota, and 6 Oneota sites, 2 in Wisconsin and 1 each in Michigan, Illinois, Kansas, and Missouri. Of the 8 remaining sites 5 are in Michigan; and there is 1 each in Indiana, Iowa, and Nebraska (Map 4).

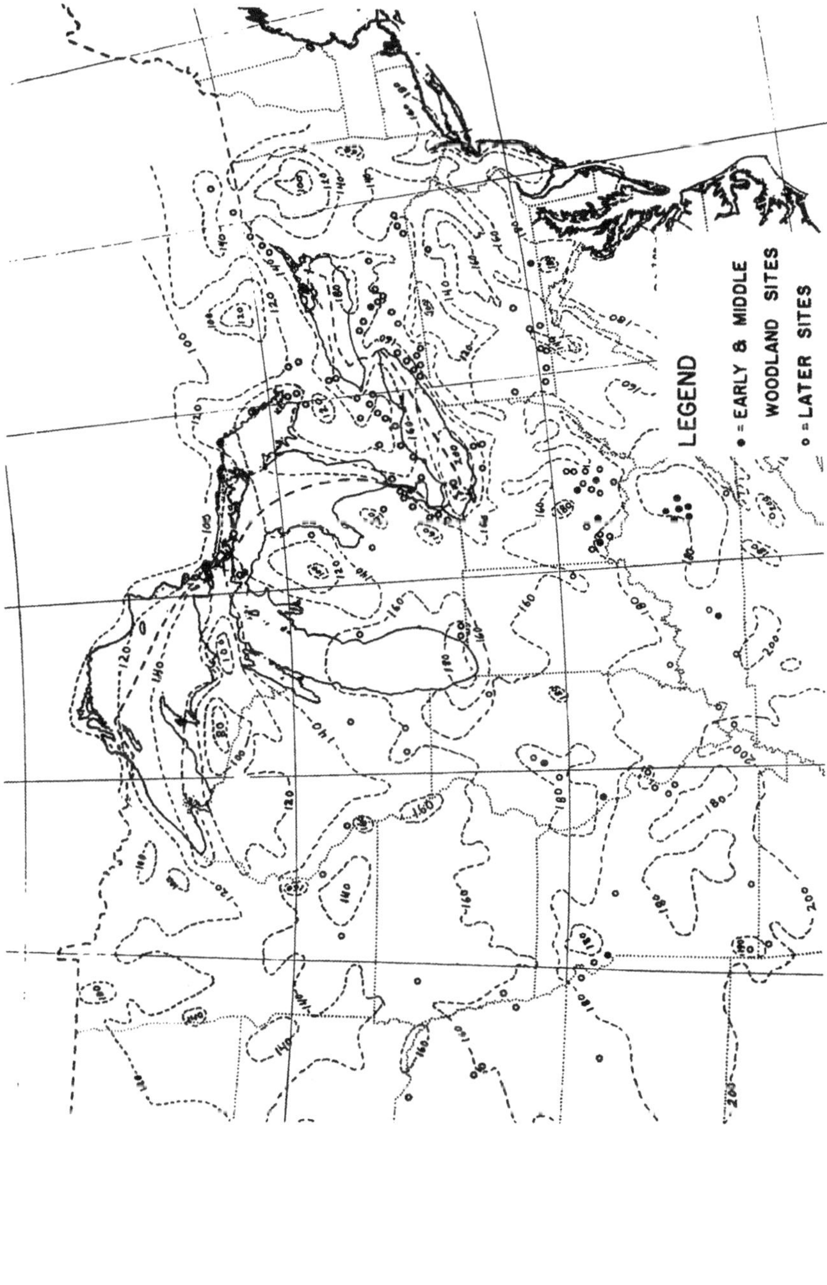

PREHISTORIC SITES WITH CULTIGEN REMAINS AND LENGTH OF FROSTLESS SEASON IN THE MIDWEST.

Map 4.

Many statements have been made about the northern limits of Indian agriculture in North America and the climatic factors involved, especially the frost-free season. Driver and Massey (1957: 215) state that corn was grown at north latitude 47 degrees in North Dakota and in the lower St. Lawrence Valley. Weatherwax (1954: 51) cites a limit of 47 degrees and 30 minutes in North Dakota and states (p. 48) that some varieties "matured in localities where the season free of frost was scarcely more than two months in length. . . ."

Mangelsdorf and Reeves (1939: 8) point out that the modern Gaspé Flint from the Gaspé Peninsula in Canada matures in 60 to 70 days. Volney H. Jones (personal communication, 1962) grew the Gaspé Flint in Ann Arbor, Michigan, where it reached maturity in 74 days but reached the milk (green corn) stage in 55 to 60 days. Nevertheless, the average frost-free period is more than 100 days in the area from which this corn was received (*Atlas of Canada* [Canada, Dept. of Mines, 1957] Plate 23). This is at 48 degrees north latitude. Reports are about the same for the maturing time of the native corn from the upper Missouri Valley in North Dakota where the average frost-free period is 120 days or more.

Byers (1946: 11) states that the limit of 100 frost-free days "marks the northern limit of commercial agriculture." The limit of 100 days for commercial corn-growing would seem to be too low except in areas where average growing season temperatures are rather high.

We have found no early historic reports of corn growing north of 47.5 degrees nor of corn growing in an area with an average frostless season of less than 120 days. In addition to this we find that prehistoric archaeological sites producing the remains of crop plants are all located south of 47.6 degrees and within the 120-day limit. This strongly indicates that one of the effective limits of prehistoric Indian agriculture was the line marking the limits of the 120-day average frost-free period, even though the Indians had corn that could mature in 60 to 70 days.

There appear to be several reasons for this. The Indians could not predict the beginning nor end of the frost-free season. If they were unlucky enough to plant too early, their crops were killed by frost; and they had to plant again. If they planted too late they were in danger of losing their crops to an early fall frost. Of course, they knew from experience the approximate time at which they should plant by observing certain springtime biological phenomena in their surroundings. Planting time had to succeed the date at which the latest frost occurred in most of the years. This would be well after the average date of the latest frost.

Growing 70-day corn where the average frost-free period is 120 days allows about 25 days to spare at either end of the growing period. This is more than an adequate margin of safety for the average year but not for the years when there are late spring or early fall frosts. According to the *1941 Yearbook of Agriculture*, the mean frost-free period is 114 days at Grayling, Michigan; but the shortest such period on record for the years of 1899 to 1938 is 72 days; and the difference between the latest spring frost and the earliest fall frost is only 48 days. In Upper Michigan the figures are almost the same for Newberry. At Stambaugh, however, where the mean frostless season is 94 days for the same years, the shortest season recorded is 42 days; and the difference between latest spring and earliest fall frosts is only one day. Twenty-four per cent of the seasons are less than 70 days in length, and 34 per cent are less than 80 days in this western Upper Michigan area.

At Winnibigoshish Dam in Minnesota where the average frost-free period is 129 days, the shortest season is 90 days; and the latest spring and earliest fall frosts are 67 days apart. These figures are 120 days, 62 days, and 49 days for Dickinson, North Dakota, with only 2 years out of 38 below 70 days. For Angelica and Indian Lake, New York, they are 121, 80, and 58 days and 84, 37, and 14 days respectively. At Indian Lake 35 per cent of the 37 years reported had frostless seasons of less than 70 days.

The data cited above indicate that the Indians who lived in areas where the frost-free period was less than 120 days could not depend upon maize agriculture for subsistence, nor could they do so in areas with longer frost-free periods where other climatic factors and soil conditions or physiography were not suitable.

L. L. Zook (1915: 7-8, 1-3) discusses some of these factors:

> The time required for any variety of corn to reach maturity depends largely upon the conditions of temperature, sunshine, and moisture of the locality where grown. The performance of a variety in one locality cannot, therefore, be taken as an indication, except in a general way, of the time required by that variety to mature in other localities where conditions of growth are different.
>
> . . .
>
> The extreme range in time of maturity [for each variety] is about 25 days. The earliest varieties [grown on the north-central Great Plains]

require under average conditions from 75 to 85 days, while the latest ones require from 100 to 110 days to mature.

. . .

The chief limitation to corn production on the Great Plains is that of climate. This area is characterized by scant and uncertain precipitation and short, variable growing seasons. The effect of these influences vary in amount from only slight injury to total crop failure and in extent from restricted localities to extended areas. A year seldom occurs in which as a result of these influences there is not some restriction to the growth and development of the corn plant.

. . .

In the most northern sections and in the high altitudes along the western border, the short season prevents the growing of any except very early varieties of corn, and these reach full maturity only in favorable years.

It is interesting in view of the above that the northwesternmost limit of the upper Missouri village agriculturalists appears not to have been determined primarily by length of frost-free period. George F. Will (1924: Fig. 14) has mapped the distribution of many upper Missouri archaeological sites. They terminate abruptly at the mouth of the Little Missouri River. This is the northwestern limit of 12 inches of warm season precipitation, and it is the lowest point at which the isotherm of 68 degrees average July temperature crosses the Missouri River. The average frostless season, however, increases to 130 days farther up the river.

It is unlikely that moisture was much of a problem to aboriginal cultivators in the Upper Great Lakes region. The average warm season precipitation is nowhere less than 15 inches, and the relatively cool summers inhibit excessive evaporation. This is generally true throughout the northern woodlands. If there is too much precipitation, growth is inhibited because of reduced sunshine and temperature; and thus the crops will take longer to mature and be in greater danger of damage by early frost. Temperatures as low as 60 degrees greatly retard flowering and maturing. The same will result, however, if there is an abundance of sunshine but not enough rain and if temperatures are so high that evaporation lowers moisture enough to delay maturity beyond the end of the frostless season. Thus, optimum conditions for early maturity as well as for high yield occur only when the various factors are in proper balance. Longer growing seasons help to compensate for other factors and vice versa.

The most important climatic factors for agriculture in the Upper Great Lakes region are those involving temperature. Of these the one most extensively considered here is frostless season, the data for which were taken from *Climate and Man* (U. S. Dept. of Agric., 1941) and the *Atlas of Canada* (Canada, Dept. of Mines, 1957). The average number of consecutive frost-free days per year was estimated as closely as possible for each archaeological site with cultigen remains. This was accomplished by comparing its location to that of the nearest lines demarcating zones of frostless season on the appropriate state map and to records for the nearest locations.

Average frostless season was computed for individual latitudes at 30 minute intervals. This was accomplished by measuring the composite length of each frostless season zone along the latitudes between 75 and 95 degrees west latitude and then calculating the average number of frost-free days for each latitude. These averages were plotted on a graph and connected by a line so that the number of days by which each site deviated from the approximate mean for its latitude could be measured.

The average frostless period for the whole region is 158 days. The average for all sites is 164 days. For the sites with corn it is 163 days, for beans 163 days, for squash 168, for gourd 173, and for sunflower 159. Sites with only Eastern Complex corn average 159 frost-free days, while those including corn of other types average 190 days.

Finch and Baker (1917: 29) state that there is very little corn grown where the mean summer temperature is less than 66 degrees Fahrenheit or the mean summer night temperature is less than 55 degrees. The region of greatest production in the United States averages 70 to 80 degrees in summer with mean summer night temperatures over 58 degrees and an average frost-free period of more than 140 days. Most strains of corn need temperatures above 50 degrees for satisfactory germination.

Judging from corn remains found in archaeological sites, prehistoric Indian agriculture did not always meet all of these requirements. Four Iroquoian sites in Ontario, the Mikado Enclosure in Alcona County, Michigan, and the Juntunen site on Bois Blanc Island are north of the modern 66 degrees summer mean isotherm. Presumably the influence of Lake Huron usually prevents low summer night temperatures at Juntunen and perhaps the others to some extent. There are records of corn from 6 Iroquoian sites and 2 Owasco sites in New York and Ontario and from the Mikado Enclosure, all apparently in areas with average frost-free seasons of less than 140 days (Table XIV, p. 138).

These are only 9 of 52 sites from which were recovered the remains of cultivated plants and which are located between 42 and 46 degrees north latitude east of the Dakotas and west of New England. Of these sites, 49 contained corn, 15 beans, 9 squash, and 6 sunflower seeds. The average of the mean frostless periods for these sites is 152 days, and the average for the whole area is about 140 days. The range in mean frostless periods for the 52 sites is from about 120 to about 170 days.

Most of the sites are in areas where the agricultural season is longer than the regional average, and the average site has a season almost two weeks longer. This indicates that there was selection of sites on the basis of frost-free period as well as on the basis of factors that often have been discussed by other writers. This selection probably resulted from agricultural experience.

By way of contrast, there are 49 archaeological sites between 37 and 41 degrees north latitude stretching from the Missouri—Mississippi Valley to the Appalachians from which cultigen remains have been recovered. This region has an average frost-free period of about 176 days which is the same as the mean of the average frost-free periods of the 49 sites that it includes. Thus, on the average, there appears to have been little selection of sites in this region on the basis of frostless season.

If we divide the entire region between southern Kentucky and the Straits of Mackinac into three subregions, each including three degrees of latitude, we find that the northern subregion averages about 138 consecutive frost-free days, the central subregion about 159 days, and the southern subregion about 181 days. The site averages in the respective subregions are about 149 days, 164 days, and 181 days. Thus the difference from north to south is 11 days, 5 days, and 0 days; and the decrease in difference from north to south is about 2 days for each degree of latitude. This decrease is not constant but is somewhat irregular as the result of various factors, primarily physiographic.

Here are some of the more obvious examples of site selection on the basis of frostless season. Hochelaga, the Huron site at which Cartier mentioned corn growing, is located at Montreal in an area where the average frost-free period exceeds 140 days. This area extends up the Ottawa River for a distance of about 120 miles, but all around it the frost-free period is less than 140 days. In fact, there are near-by areas to the north, south, east, and west that average less than 100 consecutive frostless days. Again in Ontario, the area called Huronia between Lake Simcoe and Georgian Bay has an average frostless season exceeding

140 days. This is the early historic homeland of the Huron where many village sites have been located. The frost-free period is lower to the east and south. To the southwest between Lakes Ontario and Huron there is a large area averaging less than 130 frostless days with much of it below 120 days. This area, which appears to have been unoccupied, neatly separates the territories of the Huron, Tobacco, and Neutral. The latter two territories have over 160 frost-free days in some sections, notably in the Bruce Peninsula, which was extensively cultivated.

In Lower Michigan there is a large area in the central and northeast portions with a mean frostless season of less than 140 days and diminishing to less than 90 days in the center of this area. It includes all, or most, of 18 counties and large parts of 9 others extending from the Lake Huron shore on the northeast to Montcalm and Newaygo counties on the southwest. Of 11 sites in Michigan yielding cultigen remains, only the Mikado enclosure is located within this region; and the corncob found there is quite immature. Of the 81 Indian village sites in Lower Michigan for which we know the name and location (Hinsdale, 1931: Map 2) only 3 are located within the region of less than 140 days frostless season, and none of them lies within the large territory of less than 130 days.

A strip of territory with frost-free season above 140 days runs northward along the western side of Lower Michigan terminating in the area around the Straits of Mackinac. It includes the western half of Bois Blanc Island where the Juntunen site is located. This site has produced charred kernels of corn and is the only such site that almost reaches the isotherm of 66 degrees average July temperature. It is likely that corn culture reached the Straits area after being carried up the west side of Lower Michigan, though it may have come from the Green Bay Area. There are records of fossil corn fields on Mackinac Island and on the adjacent coast of Upper Michigan (Hinsdale, 1928: 31, 34).

All of the 86 earthwork enclosures of Michigan lie south of the isotherm of 68 degrees average July temperature or south of the isotherm of 66 degrees average summer temperature, or both, except for one in Roscommon County marked "vague" in Hinsdale's *Atlas* (1931: Map 12). Only 3 of these earthworks are located north of the 68 degrees isotherm for July. They are less than 15 miles north of it and in an area of greater than 140 consecutive frost-free days. This is especially significant in view of the limitation of sites in North Dakota by the same isotherm. The implications of this are that the earthwork enclosures are associated with agriculture and presumably represent village sites. This is not to say that these isotherms are stable, but it seems unlikely that they have patterns radically different from what they were several hundred years ago. Radiocarbon dates for 3 of the northernmost enclosures are 750, 600, and 550 years ago. Most of the corn from the Juntunen site apparently dates from the same period.

Roughly, the southwestern third of Lower Michigan has a frostless season exceeding 150 days with extensions up the west side almost to Benzie County, through the central portion to Saginaw Bay, and in the southeast to Lake Huron. Nine of 11 Michigan sites with cultigen remains are within this zone. Particularly interesting is the narrow extension through Saginaw County which includes most of the Saginaw Basin with its rich alluvial soil where the greatest concentration of archaeological sites in Michigan is located and where Hopewellian pottery has been found (Quimby, 1941: 132).

Finally, there is an area around the southern half of Lake Michigan with a frostless season exceeding 160 days (much of it exceeding 170 days) with extensions up the St. Joseph and Grand River Valleys. "The east shores are more warmed by the lakes than the west owing to the prevailing westerly winds" (Kendrew, 1942: 316). It is in this area that the components of the Hopewellian Goodall Focus are located. Of the 10 sites described by Quimby (1941), 2 are within the 150-160-day zone, 3 are within the 160-170-day zone, and 5 are within the 170-180-day zone. Eight of the sites are south of the isotherm of 70 degree mean July temperature, and the other 2 are just north of it. The time factor involved, however, is on the order of 1500 to 2000 years; and climatic patterns and intensities may have been somewhat different at that time. Still it is likely that an area of relatively long frostless season existed around the southern half of Lake Michigan then as now.

In Wisconsin there are 5 sites which have produced cultigen remains, and all are located within the zone with frostless season greater than 150 days. This zone exists as a narrow strip extending up the Mississippi River to the Minneapolis—St. Paul area and a relatively small area in southeast Wisconsin with a strip adjacent to Lake Michigan extending northward to Green Bay. The Dietz, Carcajou Point, and Aztalan sites are situated in a rather isolated zone in southeast Wisconsin that has a frostless season exceeding 160 days. The Dietz site is located within a restricted zone exceeding 170 days and surrounded by territory of less than 170 days frostless season.

The other 2 Wisconsin sites producing cultigen remains are McCauley and Shrake-Gillies. The former is near Lake Winnebago, the latter is near the Mississippi River, and both lie within narrow northward projections of zones with greater than 150 consecutive frostless days. The Cannon Junction site in Minnesota is located farther up the Mississippi River from Shrake-Gillies but still within the greater than 150-day zone.

Also in Wisconsin there appears to be a definite coincidence of the lines demarcating frostless season zones and tribal territorial boundaries in early historic times as depicted by Kroeber (1939) and by Driver and Massey (1957). The line of 140 frostless days enters Wisconsin from the east near the mouth of the Menomini River and continues in a southwesterly direction along the northwestern boundary of Winnebago lands and, at the same time, along the southeastern boundaries of the Menomini, Sauk, Fox, and Kickapoo. It then turns northwestward to follow the boundary line between Kickapoo and Ojibwa on the north and Illinois and Sioux on the south. In Minnesota the boundary between Ojibwa and Sioux is more nearly approximated by the line of 130 frost-free days. Thus, in Wisconsin, the territories of Winnebago, Illinois, and Sioux are mostly in the zone of greater than 140 frost-free days while those of Menomini, Sauk, Fox, Kickapoo, and Ojibwa are mostly in the zone of less than 140 frost-free days. In fact, most of the Menomini, Sauk, Fox, and Ojibwa territory has a frostless season of less than 130 days as does much of the Kickapoo territory.

This is not to say that the Algonkian tribes of northern Wisconsin could not have depended upon agriculture to any great extent, but it does indicate that agriculture was more precarious for them than it was for the Illinois and the Siouan groups, especially in years of late spring or early autumn frost.

Although the Winnebago territory was in the zone of greater than 140 frost-free days, it was in a region with cooler summer temperatures than those occurring in Kickapoo territory; so they probably were not in a more favorable agricultural location.

In Illinois the Kingston, Crable, and Clear Lake sites are located within an isolated zone of greater than 180 frost-free days which is completely surrounded by territory of less than 180 frost-free days. Cahokia is situated within a very small and isolated area with a frostless season exceeding 200 days. With the exception of a small section of the southern shore of Lake Erie, this is the northernmost area in the interior of North America with such a long frost-free period.

In Ohio, all 19 of the sites yielding cultigen remains are within the two zones of long frostless season. In New York, more than half of the cultigen-bearing sites are situated within the relatively small territory exceeding 160 frostless days. The New York Iroquoian sites are isolated from the Monongahela sites in southwestern Pennsylvania by a substantial zone of short frostless season, just as the Whittlesey Focus sites of northern Ohio are isolated from the Fort Ancient Aspect of southern Ohio. Except for a narrow corridor along the Mohawk River, the Iroquoian sites are similarly isolated from the New England Algonkians.

The 10 Middle Mississippian sites have a longer average frost-free period than do those of any other group with cultigen remains. These Mississippian sites range in latitude from about 37 degrees to about 43 degrees and in frost-free period from over 160 days to about 215 days. The extremes in both cases are Wyckliffe and Aztalan. The average frostless season for these sites is about 190 days. This is about 16 days longer than the average for the latitudes at which these sites are located and about 23 days longer than the average of all other cultigen-bearing sites within these latitudes in the area under study.

The 8 Upper Mississippian sites (excluding the Fort Ancient Aspect) have frostless seasons averaging 168 days and ranging from about 150 days to 190 days. They range in latitude from about 39 degrees to 44.5 degrees and exceed the average frostless season of the latitudes at which they occur by about 16 days.

The 10 Fort Ancient sites have a latitudinal range of only 1.5 degrees and a frostless season range of only 170 to 190 days. The average of their frostless seasons is 178 days which is only 3 days above the average for their latitudes. The latitudinal range of the 5 Hopewellian sites is the same as that of the Fort Ancient sites, but it is about one-half degree farther north. Yet the average frost-free period for the sites is 182 days (ranging from 173 to 193 days) which exceeds the average for their latitudes by about 10 days.

Thirty-seven Iroquoian sites from which the remains of cultivated plants have been recovered range from about 120 to over 200 consecutive frost-free days and from 41.3 to 45.5 degrees north latitude. They average 153 frostless days which is seven days above the average for their latitudes. Nine Owasco sites between 41.7 and 42.9 degrees range from 138 to 168 frost-free days and average 154 frost-free days, but this is only one day above the latitudinal averages. Twelve New York Iroquoian sites average 161 days which is 9 days above the latitudinal average, and 6 New York Owasco sites average 156 days which is 3 days above the average for their latitudes. If the differences between the data for the Iroquoian and the Owasco sites are not fortuitous, it would seem either that the Iroquois were more sophisticated in site selection or that climatic patterns were not the same during the two periods of occupation.

The Monongahela sites in southwestern Pennsylvania are the most divergent from their latitudinal averages of any in the region under consideration. At 148 frost-free days, they fall 22 days below the average for their latitudes though they range from 140 to 165 days. Five of these sites are Late Monongahela while 1, at 150 days, is Early Monongahela.

Also falling below their latitudinal averages are the 14 cave sites of Kentucky, Ohio, and Pennsylvania. Though they average 174 consecutive days per year without frost, they are 5 days below the latitudinal mean. Most of these caves are rather early among agricultural sites, however; and 6 of them did not produce corn.

There are certain chronological implications of the agricultural site distribution relative to frostless season. Of the sites which may represent Early Woodland agriculture, none is within present-day zones of less than 165 frost-free days. Hopewell sites that have produced the remains of cultivated plants do not fall below an average of 170 frost-free days. Up to about 1000 A.D., 140 consecutive frostless days appears to have been the limit, whereas some early historic and protohistoric sites apparently approached 120 frost-free days.

Assuming that agriculture reached the Upper Great Lakes region during the Hopewell invasion of southwestern Michigan, it seems unlikely that this region was ever again totally without agriculture; because the territory around the lower St. Joseph River averages above 180 frost-free days at the present time.

TABLE XIV

ARCHAEOLOGICAL SITES AND AVERAGE FROSTLESS SEASON

Site	Culture	State	Mean Days Frostless Season	Deviation From Lat. Mean +	Deviation From Lat. Mean -
Harlan Rock Shelter	.....	Ky.	180	..	10
Newt Kash Hollow	(Adena?)	Ky.	178	..	5
Salts Cave	.....	Ky.	185	..	0
Steven De Hart Cave	(Adena?)	Ky.	180	..	5
Mammoth Cave	.....	Ky.	185	..	0
Little Ash Cave	(Adena?)	Ky.	178	..	5
Dillard Stamper Cave	(Adena?)	Ky.	177	..	5
Worth Creech Cave	(Adena?)	Ky.	177	..	5
Red Eye Hollow	(Adena?)	Ky.	178	..	5
Canter's Caves	.....	Ohio	170	..	5
Ash Cave	.....	Ohio	160	..	10
Kettle Hill Cave	.....	Ohio	169	..	0
Florence Mound	Adena	Ohio	170	..	0
Cowan Creek Mound	Adena	Ohio	165	..	10
Renner	Hopewell	Mo.	193	20	..
Ansell	(Hopewell?)	Ill.	175	..	0
Apple Creek	Hopewell	Ill.	180	5	..
Clear Lake	(Hopewell?)	Ill.	183	15	..
Turner	Hopewell	Ohio	190	15	..
Harness	Hopewell	Ohio	173	..	0
Wray	(Pt. Peninsula?)	N.Y.	161	5	..
Rock Creek	(Vinette I?)	Pa.	170	..	0
Walker-Gilmore	Sterns Creek	Neb.	170	10	..
Lawson	Loseke Creek	Neb.	160	5	..
Varner	Early Monongahela	Pa.	150	..	20
Steed-Kisker	Middle Mississippi	Mo.	192	20	..
Cahokia	Middle Mississippi	Ill.	200	20	..
Harrell	Middle Mississippi	Mo.	195	10	..
Long	Middle Mississippi	Mo.	193	10	..
Boyce	Middle Mississippi	Mo.	192	10	..
Aztalan	Middle Mississippi	Wis.	161	10	..
Dietz	Middle Mississippi	Wis.	170	20	..
Davis	Middle Mississippi	Texas	192	20	..
Phipps	Mill Creek	Iowa	150	..	0
Angel	Middle Mississippi	Ind.	192	10	..
Kingston	Middle Mississippi	Ill.	185	20	..
Crable	Middle Mississippi	Ill.	180	10	..
Kincaid	Middle Mississippi	Ill.	200	15	..
Wickliffe	Middle Mississippi	Ky.	215	20	..
Fisher	.....	Ill.	160	..	0
Moccasin Bluff	.....	Mich.	184	30	..
McCauley	Oneota	Wis.	156	15	..
Shrake-Gillies	Oneota	Wis.	158	15	..
Fanning	Oneota	Kans.	190	20	..
Utz	Oneota	Mo.	190	15	..
Silvernale	Oneota	Minn.	160	25	..
Cambria	Cambria	Minn.	150	10	..
Lynch	.....	Neb.	145	10	..
Kullbom	.....	Iowa	170	15	..
Larsen	.....	Neb.	160	5	..
"Ratcliffe"	.....	Ind.	155	..	15
Baum	Fort Ancient	Ohio	171	5	..
Gartner	Fort Ancient	Ohio	175	..	0
Cramer	Fort Ancient	Ohio	175	..	0
Baldwin	Fort Ancient	Ohio	169	..	0
Feurt	Fort Ancient	Ohio	180	..	0
Anderson	Fort Ancient	Ohio	173	..	0

TABLE XIV *(CONT.)*

Site	Culture	State	Mean Days Frostless Season	Deviation From Lat. Mean	
				+	-
Campbell Island	Fort Ancient	Ohio	175	..	0
Madisonville	Fort Ancient	Ohio	190	15	..
Turpin	Fort Ancient	Ohio	190	15	..
Fox Farm	Fort Ancient	Ky.	181	..	0
Juntunen	.....	Mich.	141	20	..
Mikado	.....	Mich.	125	..	10
Martin	Late Monongahela	Pa.	165	..	5
Phillips	Late Monongahela	Pa.	140	..	30
Reckner	Late Monongahela	Pa.	145	..	25
Powell I	Late Monongahela	Pa.	145	..	25
Emerick	Late Monongahela	Pa.	145	..	25
Buchanan Shelter	.....	Pa.	160	..	5
Sheep Rock Shelter	.....	Pa.	150	..	15
Lyons Farm	Owasco	Pa.	145	..	10
Castle Creek	Owasco	N.Y.	161	10	..
Bates	Owasco	N.Y.	152	..	0
Bainbridge	Owasco	N.Y.	139	..	10
Owasco Lake	Owasco	N.Y.	168	15	..
Sackett	Owasco	N.Y.	160	5	..
Kahler	Owasco	N.Y.	155	..	0
Goessens	Owasco	Ont.	138	..	15
Krieger	Owasco	Ont.	165	15	..
Stafford	.....	Ont.	140	..	15
"Mayville"	Iroquoian?	N.Y.	158	5	..
"Ellington"	Iroquoian?	N.Y.	145	..	5
"Fredonia"	Iroquoian?	N.Y.	171	20	..
McCullough	Iroquoian	N.Y.	148	..	5
Westfield	Iroquoian	N.Y.	173	20	..
Ripley	Iroquoian	N.Y.	170	20	..
Silverheels	Iroquoian	N.Y.	170	15	..
"Portage"	Iroquoian	N.Y.	145	..	5
"Lima"	Iroquoian	N.Y.	163	10	..
Richmond Mills	Iroquoian	N.Y.	163	10	..
Alhart	Iroquoian	N.Y.	161	10	..
Wilkins	Iroquoian	N.Y.	161	10	..
Hochelaga	Iroquoian	Que.	142	20	..
"Merkley"	Iroquoian	Ont.	135	10	..
"McLennan"	Iroquoian	Ont.	140	10	..
Roebuck	Iroquoian	Ont.	150	15	..
"Reade"	Iroquoian	Ont.	150	15	..
Dutton's Ash Pit	Iroquoian	Ont.	..	..	..
"Corsons"	Iroquoian	Ont.	120	..	15
"Fenelon" #23	Iroquoian	Ont.	130	..	10
Flanagan	Iroquoian	Ont.	142	10	..
Sidey-Mackay	Iroquoian	Ont.	130	..	10
Seed	Iroquoian	Ont.	145	..	0
Lawson	Iroquoian	Ont.	138	..	15
Coloman	Iroquoian	Ont.	140	..	10
Middleport	Iroquoian	Ont.	148	..	5
Uron	Iroquoian	Ont.	137	..	15
"Aylmer"	Iroquoian	Ont.	140	..	15
Clearville	Iroquoian?	Ont.	160	5	..
Pelee Island	.....	Ont.	205	55	..
Parker	Whittlesey?	Ont.	160	5	..
Tuttle Hill	Whittlesey	Ohio	180	25	..
South Park	Whittlesey	Ohio	175	20	..
Franks	Whittlesey	Ohio	175	20	..

TABLE XIV *(CONT.)*

Site	Culture	State	Mean Days Frostless Season	Deviation From Lat. Mean	
				+	-
Wolf	Whittlesey	Mich.	158	5	. .
20 MB 33.	. . . . .	Mich.	158	5	. .
Ford Turf Garden	. . . . .	Mich.	160	10	. .
Gibraltar	. . . . .	Mich.	165	15	. .
Younge	Owasco	Mich.	150	. .	5
Green Point	. . . . .	Mich.	158	10	. .
Great Warren Dune	. . . . .	Mich.	184	30	. .
Dumaw Creek	Late Woodland	Mich.	170	30	. .
20 MR 5	Late Woodland	Mich.	165	15	. .

VII

SUMMARY AND CONCLUSIONS

The Upper Great Lakes region is by no means homogeneous physically, biologically, or culturally. Neither is it a land of notable extremes in any of these respects. As a result of examining the geographical variation in temperature, physiography, soil, biota, language, and technology that existed aboriginally, one might infer that the only means by which the region might be classified as a unit is by reference to the lakes themselves. For instance there is no point in the entire region located as far as 100 miles from a large inland body of water, which would have certain effects on aboriginal conditions existing at that point. Such things are relative, but there is more to be said for classifying the region as a unit.

The whole region was subject, at various times in the Pleistocene, to continental glaciation and the resulting thermal, hydric, physiographic, and biotic conditions. In other words, there is some uniformity in the natural history of the region. Water conditions are much the same throughout with geographical differences in precipitation varying only slightly from the annual mean of 30 inches. Almost everywhere there are small lakes and streams. Soils over most of the land are podsol or podsolic. Most of the region is in a climatic zone classified as humid continental with cool summers. There is considerable variation in thermal conditions, but this could be expected in any interior region spanning 9 degrees of latitude. The relatively low relief over the greater part of the region, with most of the land varying no more than 1000 feet in elevation, is one factor responsible for the absence of sharp climatic gradients. This is reflected in the absence of sharp biotic gradients.

Though the vegetation of the Upper Great Lakes region is composed of four major types of forest, they tend not to be mutually exclusive, especially where climax conditions are present. In fact, most of the region is included by Dice (1938) in his *Canadian Biotic Province*, and is considered by some authors to be a transition zone between the northern coniferous and southern deciduous forests of eastern North America. Of approximately 50 major trees of the Upper Lakes region, about 30 range over most of the territory, including sugar maple, paper birch, and white cedar—the trees economically most important to the Indians. Most of the nut trees (including oaks), however, have a southerly distribution in the region, which is much the same as the aboriginal distribution of agriculture.

Of about 120 plants whose products were used for food, approximately 75 per cent range over most of the Upper Lakes region. Most of the others are confined to the more southerly parts of the region. Thus, it is likely that there was considerable uniformity throughout the region in the utilization of native plant products; and it is not unlikely that this uniformity extends backward in time for as long as man was in occupancy.

The earliest plant utilizations probably included the use of wood for tools and weapons, fibers for cordage, and berries for food. The presence of part of a carbonized chestnut in the deepest part of the Holcomb site (unassociated with chestnut wood charcoal and underlying pine, maple, beech, and oak charcoal) indicates that nuts were used for food from the beginning of the Archaic Period. If chestnut was already established in southeastern Michigan at this early date, it is likely that the flora of southeastern Michigan existed much as it was during the Woodland Period and that most of the economic plants were the same; though some of them may not yet have arrived in the area, and some may not yet have been utilized. The warming trend of 8000 to 3000 B.C. apparently was responsible for the formation of conditions which allowed the modern native flora to become established.

Butternut, hickory nut, and acorn were present in the Frontenac Island site in New York; but the only other Archaic food plant remains found near the Great Lakes are from the 4000-year-old Feeheley site in Saginaw County, Michigan. They include fragments of acorn, butternut, hickory nut, walnut, and grape seed—all plant remains that are likely to be preserved and recovered. Many others such as tubers, fruits with very small or fragile seeds, tree sap, and greens are rarely or never recovered archaeologically. Scarcity of plant remains from Archaic and even Paleo-Indian sites is just as likely due to excavation technique and decay as to lack of utilization.

Hickory nuts, walnuts, acorns, chenopod seeds, and blackberry seeds have been recovered from Early Woodland sites in Ohio and New York, but no plant remains from sites of this period have been reported for the Upper Great Lakes region. Middle Woodland sites in Ontario and northwest Pennsylvania have yielded Canada plum, hickory nut, walnut, and butternut. Walnut, butternut, and acorn were found at the Spoonville Hopewell site in Michigan. Thus, our plant food inventory for Middle Woodland is not much larger or much different from that of the Archaic.

Our evidence of plant foods is somewhat better for the Late Woodland Period. We have a list of 10 from Fort Ancient sites, 5 from Owasco sites, and 10 from Iroquoian sites. There are hickory nut, walnut, butternut, hazelnut, acorn, grape, thornapple, and plum from 4 sites in southeast Michigan. Acorn, walnut, plum, and grape occurred in 2 sites in southwest Michigan. Pond nut, hazelnut, and bearberry were recovered from 3 sites on Isle Royale.

By far the greatest amount of information about prehistoric plant foods in the Midwest comes from the Juntunen site on Bois Blanc Island which produced remains of plum, two kinds of cherry, hazelnut, bearberry, chenopod, blackberry, blueberry, grape, elderberry, beechnut, acorn, sumac berry, and pepperroot, as well as corn. The site is just north of the natural range of the other nuts. In addition there were two kinds of weed seeds *(Galium* and *Rumex)*, an unidentified tuber, an unidentified large seed, or small bulb or tuber, and abundant birchbark. The blueberries, having extremely small seeds, would never have been located had not excavation procedure preserved a cluster of three very fragile, carbonized berries. The chenopod is apparently mostly *Chenopodium album*. This is a weedy species which some authors do not include in the native flora of the region.

The catalogue of plant remains provides us not only with empirical evidence of subsistence practices but also with inferential evidence of seasonal occupation. Most of the plant foods represented at Juntunen are available some time in August (at the latest or earliest in some cases), but the minimum over-all season for collection of all these foods extends from May to October, though this might not represent any one particular year. If a carefully examined archaeological site were to produce a rather large collection of plant remains that included relatively little evidence of plant food, it would be supporting evidence for a hypothesis that the site was occupied only in winter or spring.

The ethnobotanical literature indicates that nearly 400 native plant species were utilized in one way or another by Indians of the Upper Great Lakes region, including about 130 food plants and more than 50 plants used technologically. Since the flora of the region includes approximately 2000 species, it is estimated that the Indians utilized at least 20 per cent of the kinds of plants available to them.

Eight categories of plant food have been formulated including sap and cambium, bulbs and tubers, greens, flowers and buds, fleshy fruits, seeds, nuts, and lichens. There is a tendency for the availability, or at least the collection, of each of these kinds of food to occur at a particular season. Thus we can make generalizations about seasonal subsistence activities on the basis of archaeological plant remains which can be supported by reference to ethnography. The most important plant foods by season of exploitation appear to be as follows: maple sap in early spring, tubers in spring, greens in late spring, berries in summer, wild rice in late summer, nuts in fall, tubers in late fall, and lichens in winter. In addition to the few plant foods naturally available in winter, about 40 others were stored for winter use including maple sugar, wild rice, and various berries, tubers, and nuts. Five or 10 plant foods are available in January and February. The number increases to about 25 by April and drops back to about 15 by June. In August it reaches 45 to 50, after which it drops back to 15 or 20 by December.

Perhaps the most interesting conclusion reached as a result of food plant distribution analysis is that only 12 of them are available in northern Canada and that this availability steadily increases toward the south through the Lakes region and presumably beyond. This is not unexpected, however, because the total number of plant species present increases comparably. This presumably is one factor responsible for the small proportion of plant food in the diets of the more northerly peoples. Berries are much more abundant in the north, relative to nuts and tubers which tend to have more southerly distributions.

The most important native plant foods of the historic Indians appear to have been maple sugar and wild rice, though they may have been relatively less important prehistorically. Acorns may have been more important in prehistoric than in historic times in areas where they were rather abundant. Among tubers the groundnut was important in the south and the duck potato in the north. Milkweed may have been important because it is prolific in disturbed habitats, because some edible part of the plant is available from late spring to early autumn, and because it is a source of good fiber. Chenopod may also have been important because it furnished greens in summer and seeds in autumn and early winter and because, like milkweed, it is prolific in disturbed areas and thus easily available. All of the nuts probably were important whenever and wherever they were abundant. This would be true also of the various berries, especiall the raspberry—blackberry group, the blueberry—cranberry group, the currant—gooseberry group, the Juneberries, the cherries, the grapes, and perhaps some of the thornapples.

Of the 325 native plant species utilized by the Indians for essentially nonfood purposes, 82 were also food plants. "Non-food plants" here include 18 species used for beverage and flavoring, 25 for dyeing, 27 for smoking, 31 for various supernatural purposes, and 275 for medicine. The total of 510 indicates the extent to which various plants had multiple uses.

The beverage plants, especially the nondeciduous hemlock, probably had considerable importance in maintaining vitamin levels during certain seasons, primarily in winter. This may have been equally true for the medicinal teas which seem to have been frequently taken. Actual medicines, as such, might have contributed in this way also but are less likely to have been significant to nutrition. They probably also were useful medicinally in some cases.

The medicinal uses of plants grade into the supernatural uses which include medical and ceremonial charms and incense, "medicine" for malicious magic and ceremonial dreams, sacred beads, sacred bundle components, charms for love and success, charms against snakes, and lures for hunting and fishing. It is not likely, however, that the native mind recognized the kinds of compartmentalization that are made here. A large percentage of plants in the charm category are monocots, but we have no explanation for this lest it be because of the showy flowers.

A majority of the plant products used in smoking are included in a formula smoked to attract buck deer. The odor of the smoke is said to simulate the scent carried in the hoof of the doe. Bark, flowers, leaves, root, and seeds were smoked, probably before the introduction of tobacco, which was usually only one component of a smoking mixture. Thus it is not safe to assume the possession of tobacco by a people just because pipes are part of an archaeological assemblage. Tobacco is not native to the eastern United States, the species utilized having been introduced from South America. Its present sporadic occurrence in the region is probably a relict of introduction and cultivation by the Indians.

The utilization of plants in smoking and dyeing is not likely to have been significant to Indian subsistence. Only 3 of 25 dye plants are not reported as having other uses. Thus it is probable that their use in dyeing was a secondary development in many cases.

The most useful nonfood plants were paper birch, northern white cedar, and basswood, which were rather widely and abundantly distributed through the Upper Lakes region. Birchbark was the most frequent of the plant remains recovered from the Juntunen site, with the exception of wood charcoal. Twenty-four other trees were utilized technologically, along with 7 shrubs and 18 herbaceous species. The plant parts used include bark, bast, stems, leaves, wood, roots, pitch, and thorns. The most important products were bark containers, canoes, and houses, thread, cordage, nets, bags, baskets, mats, and wooden tools and weapons.

The habitat in which a people lives is one of the factors to which their culture adapts in certain respects, that is, natural enviroment offers limits to and influences culture. But the activities of a group of people also have significant effects upon their habitat so that reciprocal influences are brought about. Some of the Indian activities which tend to alter nature include clearing and burning, collection of food and firewood, disposal of waste, and cultivation. These activities alter the soil, increase surface light and evaporation, alter natural community composition, and extend the ranges of some plants, among other things. Some habitat modifications have resulted in decreasing productivity and some have had the opposite effect as has been reported for aboriginal New England. We have little information, however, concerning the aboriginal existence of such phenomena in the Upper Great Lakes region; though it appears that the ranges of some plants were extended (especially *Nelumbo lutea,* pond nut) and that there was considerable burning, clearing, and cultivating which would be expected to have some effects. For instance, annual firing may have been responsible for perpetuation of grassland in some locations.

Climax vegetation is not very productive of many products that are significant to Indian subsistence. Maple, nut, and fruit trees are less productive in thick stands than they are in open woods. Berry bushes are found most abundantly around clearings but occur only sparsely within climax forest. The weedy economic plants, especially, need the light and disturbance of habitats such as those resulting from clearing and burning. The productivity of the Indian's cultivated plot did not end when he ceased to plant corn there, and the disclimax vegetation of his abandoned village site might be quite valuable to him on his return. Not only did in the Indian unwittingly (?) prepare habitats for many of the native economic plants, but he transported their seeds to these locations along with those of occasional exotics, which now and then gained a foothold that might last until the site reverted to subclimax vegetation, if it were allowed to do so. Most of these "camp-followers" are productive between late spring and early fall. This is generally the season during which the village sites were occupied. This general kind of situation would seem to have existed at the Juntunen site, as indicated by the remains of weedy plants and those which ordinarily grow in and around clearings.

In addition to introduction of new plant species, there is another possible effect of Indian activities on plant life in the Upper Lakes region that has persisted to the present. On the basis of random distribution one should expect the percentages of named varieties, forms, and hybrids among the Indians' economic plants to be approximately the same as their percentages among the total flora. Yet, these subspecific taxa are from two to three times as common among the economic plants as they are among the total flora of the northeastern United States and adjacent Canada. It is possible, however, that species with greater diversity and hybridization rates were more likely to be utilized. And it is possible that the plants which were of value to the Indians are more likely to be those which have been studied closely by botanists. But there does appear to be a rather strong correlation between Indian utilization and genetic diversity of native plant species.

In addition to the supposedly fortuitous preparation of habitats and introduction of economic plants as a byproduct of other activities, Indians living in the southern half of the Upper Lakes region intentionally introduced certain plants and prepared plots for their propagation. These plants included corn, squash, beans, sunflower, tobacco, perhaps gourd, and possibly one or two others. We have attempted to trace cultigen prehistory in the Midwest by examining the available archaeological evidence in the form of actual plant remains. Ethnographic evidence was avoided in order to insure that no part of the distribution was influenced by the disruptions following European contact.

The earliest cultigen remains within 300 miles of the Upper Lakes region are probably the prepottery sunflower seeds from Mammoth Cave, Kentucky, though we have no actual date. Nearby Salts Cave, which has produced a radiocarbon date of 1125 B.C. has yielded various cultigen remains, as has Newt Kash Hollow with radiocarbon dates of 690 B.C. and 640 B.C. Thus there are definite indications that plant cultivation was practiced in the Eastern Woodlands as early as 1000 B.C. and very good evidence that it was present in Early Woodland cultures.

The earliest cultigens appear to have been sunflower, marsh elder, squash, gourd, and perhaps chenopod. There is not sufficient evidence that corn was cultivated in the East before Middle Woodland times or that beans appeared before the Late Woodland period. Sunflower (and perhaps chenopod) may have declined somewhat in relative importance with the arrival of corn, and marsh elder apparently disappeared as a cultigen by the time that beans were present in the East.

Outside of Kentucky the only Early Woodland cultigen remains are gourd and perhaps squash (and chenopod?) from two Adena sites in Ohio. It is possible that some of the cultigens from Ohio and Arkansas caves represent Early Woodland cultures, but this remains to be demonstrated.

Corn has been reported from six Hopewellian sites; marsh elder was present in one and a clay effigy gourd in another. Otherwise, direct evidence of Middle Woodland cultigens is lacking (unless chenopod be included). Two of the Hopewellian corn reports are questionable, but the others appear to be quite valid. It is likely that sunflower and squash continued in cultivation throughout Middle Woodland times. It is possible that tobacco was introduced at this time or even earlier, but archaeological evidence of this plant in the East is so poor as to be almost useless for determining its time of arrival.

Hopewellian culture intruded well into the Upper Great Lakes region, especially as represented by the Goodall Focus in southwestern Michigan. Presumably corn and perhaps squash and sunflower were cultivated, but there is no direct evidence of it. No gourd or tobacco remains have been reported from any prehistoric site in the Upper Lakes region.

Unfortunately, we know very little about the kind of corn that was cultivated by the Hopewell people, but we do know that the peoples of the succeeding Middle Mississippian culture were growing a type very similar to one of the Southwest called "Basketmaker." The later Mississippians, however, seem to have abandoned this corn for a type called "Northern Flint" or "Eastern Complex" which was grown throughout the eastern United States and survived to contribute to the modern hybrid varieties.

In the Upper Lakes region only corn from the Owascoid Younge site in southeastern Michigan displays definite tendencies toward the Mississippian Basketmaker type. Otherwise it is all essentially Eastern Complex, though some of the Juntunen corn tends to deviate somewhat from the typical characteristics of that race. In any case the Younge and Juntunen corn remains appear to be the earliest definite evidence of Upper Great Lakes agriculture. Squash from Aztalan in southeastern Wisconsin and corn from the Goessens and Krieger Owasco sites in extreme southern Ontario are the nearest sources of cultigen remains that may be older. Otherwise one must look to the Illinois River Valley and southern Ohio for existing evidence of Midwest agriculture that is definitely older than the Younge and Juntunen corn.

In addition to corn, we have evidence of squash, sunflower, and beans from Middle Mississippian sites; sunflower and beans from Fort Ancient sites; beans from Owasco sites; and squash, sunflower, and beans from Iroquoian sites. In the Upper Lakes region there are corn, squash, beans, and sunflower seeds from both Flanagan and Sidey-Mackay near Georgian Bay in Ontario. In Michigan, corn has been found at 20 MR 5, Gibraltar, Wolf, and Verchave II in the Detroit area, at Moccasin Bluff and Great Warren Dune in Berrien County, at Green Point near Saginaw, and at the Mikado Earthwork in Alcona County. Beans occurred at 20 MB 33 in Macomb County and squash at Dumaw Creek in Oceana County.

Cultigen remains have been reported from at least 120 archaeological sites in the greater Midwest—that area east of the Dakotas and north of Tennessee from the lower Missouri Valley to western Pennsylvania and Montreal. None of these sites is located north of 46 degrees north latitude, whereas sites in North Dakota as far north as 47.6 degrees have produced archaeological corn.

There are corn varieties that will mature in 60 to 70 days under optimum conditions, but it appears that an average frostless season of 120 days was the limit for almost all aboriginal agriculture in the Midwest. The limit seems to have been gradually decreasing, however, as Indian crop plants were becoming adapted to shorter growing seasons. In the Upper Lakes region of northern Wisconsin, most of Upper Michigan, northeastern Lower Michigan, and all of Ontario except the zone along the western and southern shores of Lake Huron, fall below the 120-day average frost-free season.

In addition to minimum growing season, crops require adequate night temperatures, daytime temperatures, and average temperatures for certain lengths of time before they will mature. They also need enough, but not too much, moisture and sunshine. Thus, all minimum conditions must be met before a crop can mature. If average temperature is low, a longer growing season is required; so it usually takes longer to mature a particular type of corn in the north than it does in the south. Longer summer days in the north probably help to relieve the effect of temperature deficiency, however.

Perhaps the most important climatic factor for aboriginal agriculture in the Upper Lakes region was length of frostless season. Sites with cultigen remains are located almost exclusively within the zone of greater than 130 consecutive frost-free days. They are also south of the isotherm of 66 degrees average July temperature. There are 52 sites in the greater Midwest between 42 and 46 degrees north latitude which have produced cultigen remains. The average frostless season for these sites is 152 days, but the average for the general area in which they are located is only 140 days. In contrast to this, the 49 cultigen-bearing sites between 37 and 41 degrees have an average frostless season of 176 days which is the same as the average for the area occupied by them. Dividing the greater Midwest into three equal latitudinal divisions we find that the divergence in number of consecutive frost-free days for the sites and the areas is eleven, five, and zero from north to south. This decrease in divergence is almost two days for each degree of latitude.

There are a number of areas in the region that exist as pockets or extensions of relatively long-growing season. It is in these areas that cultigen-bearing sites are most common. In the Upper Lakes region these areas include Huronia, between Lake Simcoe and Georgian Bay; the Saginaw Valley; the Straits of Mackinac; and southwestern lower Michigan – the Goodall Focus territory. Five Wisconsin sites with cultigens have been reported, all of which are within extensions of the zone of greater than 150-day frostless season; and three are located within a limited extension of the 160-day zone.

It is difficult to avoid the conclusion that the agricultural sites of the southern Upper Lakes region were selected largely on the basis of duration of frostless season. This is an indication that the use of good climatic maps can aid the anthropologist in locating the boundaries of aboriginal cultures in some areas.

In summary, the following general conclusions can be drawn. Plant foods have been utilized in the Upper Great Lakes region at least since the beginning of the Archaic Period, but the data are too meager at the present time to recognize any definite patterns. Evidence from the Juntunen site, however, indicates that with the use of careful techniques of excavation and analysis the data can be greatly augmented.

There do appear to be definite patterns of plant utilization that can be discovered through synthesis of ethnographic and botanical records. When combined with the archaeological evidence, other kinds of information can be produced.

Evidence of the influence of Indian activities on their habitats is difficult to collect and interpret, but with the combined exploitation of various empirical resources of anthropology and biology new insights into aboriginal ecology are possible.

By plotting the time and space distributions of the archaeological remains of cultigens in relatively restricted areas, we can gain more refined knowledge of the history of agricultural dispersal and development. And, finally, by comparing these distributions to those of various natural factors we can add new evidence for the interpretation of problems in aboriginal demography.

APPENDIX A

BEVERAGE AND FLAVORING PLANTS

In the following list of eighteen plants, each entry includes the scientific and common names of the plant, followed by the part utilized and its season of availability, the tribes recorded as using it, other uses, and the range and habitat preferences of the plant. Occasionally other comments are included. The order of the entries is taxonomic, following *Gray's Manual of Botany* (Fernald, 1950). Six of these plants were used for seasoning foods and sixteen for beverages as noted below.

1. *Tsuga canadensis* (L.) Carr. Hemlock.

Leaves available anytime. Used by Ojibwa for beverage (Smith, 1932: 408; Bakeless, 1950: 115). Also used for medicine, dye, and utility. Grows from southern lower Michigan to Lake Superior on well-drained uplands and ravine slopes. This is a major element of the climax forest of the Upper Great Lakes region.

2. *Comptonia peregrina* (L.) Coult. Sweet fern.

Leaves available from late spring to autumn. Used by Ojibwa, Menomini, and Potawatomi for beverage (Gilmore, 1933: 127; Smith, 1923: 42; 1933: 65). Also used for medicine and ceremonial incense. Grows from Ohio to southern Canada in open, sterile woodlands, clearings, and pastures.

3. *Quercus ellipsoidalis* E. J. Hill. Jack oak.

Acorns available in autumn. Ground for making a beverage by Menomini (Smith, 1923: 66). Grows from northwestern Ohio to western Upper Michigan and Minnesota in well-drained uplands.

4. *Morus rubra* L. Red mulberry.

Berries available in summer (a good crop every two-to-three years). Used by the Iroquois to make a summer drink (Parker, 1910: 96) and for food (Waugh, 1916: 128). Grows from the Gulf of Mexico to central Lower Michigan on rich soil in river bottoms.

5. *Asarum canadense* L. Wild ginger.

Roots available in spring and summer. Used by Ojibwa, Potawatomi, and Sauk-Fox for flavoring (Smith, 1932: 397; 1933: 41; 1928: 255). Also used for food and medicine but not reported for beverage. Grows from Missouri and North Carolina to Minnesota and Quebec in rich woods and shaded calcareous ledges.

6. *Lindera benzoin* (L.) Blume. Spice bush, Benjamin bush.
Leaves available from spring to autumn; twigs available anytime. Used by Chippewa for flavoring and beverage (Gilmore, 1933: 131). Also used for medicine. Grows from Kentucky to southern Michigan in damp woods and brooksides.

7. *Sassafras albidum* (Nutt.) Nees. Sassafras.
Roots available anytime. Used by Iroquois and Ojibwa for flavoring and beverage (Harris, 1891; Gilmore, 1933: 131). Also used for medicine. Grows from the Gulf of Mexico to central Lower Michigan in well-drained, stony or sandy soil, woods, abandoned fields, and peaty swamps.

8. *Hamamelis virginiana* L. Witch hazel.
Leaves available in summer. Used by Iroquois for beverage (Fernald and Kinsey, 1943: 228). Also used for medicine and sacred beads. Grows from Tennessee to Minnesota and southern Quebec (common throughout Michigan) in woods and borders of woods.

9. *Rhus typhina* L. Staghorn sumac.
Berries available from midsummer to early winter. Used by Ojibwa, Menomini, and Potawatomi for beverage (Smith, 1932: 397; 1933: 38; 1923: 22). Dried for winter by Menomini. Also used for food, dye, and medicine and smoked with tobacco by the Potawatomi. Grows from Kentucky to southern Ontario and Minnesota (common throughout Michigan) in dry or gravelly soil, especially on open ridges. Seeds of this or the next species found archaeologically at the Juntunen site on Bois Blanc Island, Michigan.

10. *Rhus glabra* L. Smooth sumac.
Berries available in summer and autumn. Used by Ojibwa for beverage (dried for winter) (Smith, 1932: 397). Also used for medicine and dye and smoked by the Ojibwa and Potawatomi. Grows from Louisiana and Florida to Minnesota and Nova Scotia (common throughout Michigan) in dry soil, old fields, roadsides, and borders of woods.

11. *Cornus racemosa* Lam. Panicled dogwood.
Bark available anytime. Used by Ojibwa for beverage (Smith, 1932: 367). Also used for smoking. Grows from Kentucky to Maine and Minnesota (common in lower Michigan) in dry to moist open habitats.

12. *Ledum groenlandicum* Oeder. Labrador tea.

Leaves available in spring to autumn. Used by Ojibwa and Potawatomi for beverage (Smith, 1932: 401; 1933: 99). Also used for dye and medicine. Grows from Ohio to Alaska and Greenland (common in Upper and northern Lower Michigan) in bogs and damp thickets.

13. *Andromeda glaucophylla* Link. Bog rosemary.

Leaves available in spring to autumn. Used by Ojibwa for beverage (Smith, 1932: 400). Grows from Ohio to Labrador and Manitoba (frequent throughout Michigan) in bogs and wet shores.

14. *Chamaedaphne calyculata* var. *angustifolia* (Ait.) Rehd. Leather leaf, cassandra.

Leaves available most of the year. Used by Ojibwa for beverage (dried for future use) (Smith, 1932: 400). Also used for medicine. Grows from Ohio to Newfoundland and Alaska (common throughout Michigan) in swamps and boggy meadows.

15. *Gaultheria procumbens* L. Teaberry, wintergreen.

Leaves available spring to fall. Used by Ojibwa and Potawatomi for beverage (Smith, 1932: 400; 1933: 56). Also used for flavoring and medicine; berry also used for food. Grows from Kentucky to central Ontario (common throughout Michigan) in thin sandy woods and clearings.

16. *Gaultheria hispidula* (L.) Bigel. Moxie plum, creeping snowberry.

Leaves available in spring to fall. Used by Ojibwa for beverage (Densmore, 1928: 317). Berries also used for food. Grows from southern Michigan to Hudson Bay, in bogs in the south and in mossy woods in the north.

17. *Pycnanthemum virginianum* (L.) Durand and Jackson. Basil, mountain mint.

Flowers and buds available from July to September for flavoring. Used by Ojibwa (Densmore, 1929: 39). Also used for food and medicine, but not reported for beverage (Densmore, 1928: 290, 307). Grows from Ohio to Maine and North Dakota in gravelly shores, meadows, dry-to-wet thickets, etc.

18. *Mentha arvensis* L. Mint.

Leaves available in spring to fall. Used by Ojibwa, Menomini, and Potawatomi for beverage (Smith, 1932: 405; 1923: 39; 1933: 61). Used for flavoring and medicine. Grows from Ohio to Newfoundland and Alaska mostly on gravels.

APPENDIX B

MEDICINAL BEVERAGE PLANTS

In the following list of sixty-eight plants, each entry includes the scientific and common names of the plant, plant part used, tribes reported as using the plant, other uses, and the range and habitat preferences of the plant. The order of the entries is taxonomic, following *Gray's Manual of Botany* (Fernald, 1950).

1. *Equisetum arvense* L. Scouring rush, field horsetail.
Used by Ojibwa and Potawatomi (Smith, 1932: 368; 1933: 55). Ojibwa also used it as a charm. Grows from New York, Michigan, and Wisconsin to Greenland and Alaska in open low ground, fields, sterile meadows, roadsides, damp open woods, and thickets; generally common.

2. *Equisetum pratense* Ehrh. Meadow horsetail.
Used by Ojibwa (Reagan, 1928: 248). Ojibwa also used the tuber for food *(ibid.)*. Grows from southern Michigan to central Canada in alluvial woods, thickets, and calcareous meadows.

3. *Equisetum sylvaticum* L. Wood horsetail.
Used by Ojibwa (Smith, 1932: 368). Grows from Ohio to Greenland and Alaska in woodlands, thickets, and openings.

4. *Dryopteris cristata* (L.) Gray. Shield fern, crested wood fern.
Root used by the Ojibwa (Smith, 1932: 381). Grows from Tennessee to Newfoundland and Alberta in boggy or swampy open ground and thickets or wet woods.

5. *Athyrium filix-femina* (L.) Roth. Lady fern.
Root used by Ojibwa and Potawatomi (Densmore, 1928: 287; Smith, 1933: 73). Grows from Ohio to Newfoundland and Alaska in damp thickets, meadows, swamps, and brooksides.

6. *Adiantum pedatum* L. Maidenhair fern.
Root used by Potawatomi (Smith, 1933: 73). Grows from Alabama to Quebec and Minnesota in rich hardwoods.

7. *Pteridium aquilinum* (L.) Kuhn. Bracken fern.
Root used by Ojibwa and Menomini (Smith, 1932: 382; 1923: 47). Young sprouts used in soup by Ojibwa (Smith, 1932: 408).

Grows from Mexico to Ontario and Alaska in dry open woods, pastures, recently burned clearings, etc.

8. *Taxus canadensis* Marsh. American yew.
Used by Ojibwa and Potawatomi (Gilmore, 1933: 123; Smith, 1933: 84). Menomini also used it for medicine. Berries are edible. Grows from Ohio to Newfoundland and Manitoba in rich woods and thickets.

9. *Abies balsamea* (L.) Mill. Balsam fir.
Used by Menomini (Smith, 1923: 45), Ojibwa, and Potawatomi for medicine (Smith, 1932: 378; 1933: 68); Ojibwa and Algonkin as utility plant. Grows from southern Michigan to Hudson Bay in cool, moist, rich soil and low swampy ground in the south and on well-drained hillsides in the north.

10. *Picea glauca* (Moench) Voss. White spruce.
Used by Menomini and Ojibwa (Smith, 1923: 45; 1932: 379). Grows from central lower Michigan to northern Canada in low damp woods in good soils.

11. *Larix laricina* (Du Roi) K. Koch. Eastern larch, tamarack.
Bark used by Potawatomi, Menomini, and Ojibwa (Smith, 1933: 69; 1932: 378). Ojibwa and Algonkin used it as a utility plant. Grows from northern Ohio to Labrador and Alaska, mostly in swamps in the Great Lakes region.

12. *Pinus strobus* L. White pine.
Used by Menomini (Smith, 1923: 45). Ojibwa and Potawatomi used it for medicine (Smith, 1932: 379; 1933: 70) and utility; Ojibwa and Iroquois for food. Grows from southern Michigan to James Bay in light fertile loam and sandy soil.

13. *Thuja occidentalis* L. Arborvitae, northern white cedar.
Used by Ojibwa and Potawatomi (Smith, 1932: 380; 1933: 70). Ojibwa also used it for dye, food, and as a ceremonial incense. Ojibwa and Menomini utility plant. Grows from southern Michigan to James Bay (common near the Juntunen site) in swamps and cool rocky banks.

14. *Trillium grandiflorum* (Michx.) Salisb. Trillium.
Used by Menomini, Potawatomi, and Ojibwa (Smith, 1923: 41; 1933: 63; Densmore, 1928: 294). Greens are edible. Grows from Georgia to Quebec and Minnesota in rich woods and thickets.

15. *Populus grandidentata* Michx. Big-toothed aspen.
Root used by Ojibwa (Smith, 1932: 387), cambium for food. Grows from Tennessee to southern Ontario and throughout the

Great Lakes region in rich, moist, sandy soil, swamp borders, river banks, and hillsides.

16. *Ulmus rubra* Muhl. Slippery elm, red elm.
Cambium used by Menomini (Smith, 1923: 56); by Ojibwa and Potawatomi, for medicine (Smith, 1932: 392; 1933: 86), and utility plant; cambium used by Iroquois for food. Grows from Alabama to southern Ontario and northern Wisconsin in bottom lands, stream banks, rich moist hillsides, and rocky ridges and slopes.

17. *Humulus lupulus* L. Hop.
Used by Ojibwa (Smith, 1932: 391). Grows from Ohio to New Brunswick and Montana in thickets in alluvium.

18. *Urtica gracilis* Ait. Nettle.
Used by Potawatomi (Smith, 1933: 87) and Ojibwa for medicine (Smith, 1932: 392). Ojibwa, Potawatomi, Menomini, and Sauk-Fox used it for utility fiber. Grows from Texas to Hudson Bay in thickets and rich damp soil.

19. *Laportea canadensis* (L.) Wedd. Wood nettle.
Root used by Ojibwa (Smith, 1932: 391). Ojibwa, Menomini, and Sauk-Fox used it for utility fiber. Grows from Alabama to Quebec and Manitoba in low woods and stream banks.

20. *Polygonum coccineum* Muhl. Smartweed.
Used by Ojibwa (Smith, 1932: 381). Also smoked by Ojibwa to attract deer. Grows from Mexico to Quebec and Washington in shores and margins of ponds or streams, wet prairies, etc.

21. *Polygonum pennsylvanicum* L. Heart seed.
Used by Menomini and Ojibwa (Smith, 1923: 47; Gilmore, 1933: 129). Grows from Florida and Texas to Quebec and Minnesota in damp shores, thickets, clearings, and disturbed or cultivated soil.

22. *Polygonum lapthifolium* L. Smartweed.
Used by Potawatomi (Smith, 1933: 72). Grows from Newfoundland to British Columbia and southward in swampy thickets, shores, damp clearings, and cultivated fields.

23. *Hepatica americana* (DC) Ker. Liverleaf.
Used by Potawatomi and Ojibwa (Smith, 1933: 75; Gilmore, 1933: 129). Potawatomi also used it as a dye. Grows from Alabama to Nova Scotia and Manitoba in dry woods.

24. *Anemone cylindrica* Gray. Thimbleweed.
Root used by Ojibwa (Smith, 1932: 383). Grows from Ohio to Maine and Alberta in dry open soil, prairies, and slopes.

25. *Actaea rubra* (Ait.) Willd. Red baneberry, snakeberry.
Root used by Ojibwa and Potawatomi (Smith, 1932: 382; 1933: 74). Grows from Ohio to Labrador and British Columbia in woods and thickets.

26. *Physocarpus opulifolius* (L.) Maxim. Ninebark.
Used by Menomini (Smith, 1923: 49). Grows from Tennessee and northern Illinois to Hudson Bay in shores, rocky banks, and thickets.

27. *Spiraea tomentosa* L. Steeplebush, hardhack.
Used by Ojibwa (Smith, 1932: 386). Grows from Tennessee to Quebec and Manitoba in sterile low grounds and pastures.

28. *Pyrus melanocarpa* (Michx.) Willd. Black chokeberry.
Berries used by Potawatomi (Smith, 1933: 76). Potawatomi also used it as food (Smith, 1933: 107). Grows from Tennessee to Ontario and Minnesota in swamps and low grounds.

29. *Rubus allegheniensis* Porter. Highbush, blackberry, sow-teat blackberry.
Root used by Ojibwa (Smith, 1932: 385) and Mascouten (Smith, 1933: 108) for medicine. Berries also used for food. Grows from Kentucky to northern Lake Superior in dry clearings and thickets.

30. *Prunus pensylvanica* L. Firecherry, pin cherry.
Bark used by Ojibwa and Potawatomi (Smith, 1932: 385; 1933: 77). Ojibwa, Potawatomi and Iroquois used the berry for food. Grows from northern Illinois to James Bay in sand-lands, burns, roadsides, and clearings.

31. *Prunus virginiana* L. Chokecherry.
Used by Menomini, Ojibwa, and Potawatomi (Smith, 1923: 50; 1932: 385; 1933: 78). Ojibwa, Menomini, Potawatomi, Mascouten, Sauk-Fox, Winnebago, and Iroquois used the berry for food. Grows from Ohio to James Bay along fence lines and edges of woods.

32. *Prunus serotina* Ehrh. Black cherry.
Used by Ojibwa (Smith, 1932: 385). Menomini, and Potawatomi as medicine (Smith, 1923: 71; 1933: 77). Ojibwa, Menomini, and Potawatomi also used the berry for food. Grows from the Gulf of Mexico to southern Ontario and Isle Royale in dry woods and fence rows.

33. *Zanthoxylum americanum* Mill. Northern prickly ash, Toothache tree.
Bark used by Menomini and Ojibwa (Smith, 1923: 51; 1932: 387). Menomini and Potawatomi also used the root for medicine

(Smith, 1923: 51; 1933: 80). Grows from Alabama to Quebec and North Dakota in rich woods and river banks.

34. *Polygala senega* L. Milkwort, Seneca snakeroot.
Used by Ojibwa (Densmore, 1928: 291). Also used by Ojibwa as a charm. Menomini and Iroquois used it as food. Grows from Georgia to Alberta and Quebec in dry, rocky, or gravelly areas.

35. *Nemopanthus mucronata* (L.) Trel. Mountain holly, catberry.
Used by Potawatomi (Smith, 1933: 39). Potawatomi also used the berry for food; Ojibwa used it for medicine (Smith, 1932: 355). Grows from Ohio to central Ontario in deep cedar and tamarack bogs.

36. *Ceanothus americanus* L. New Jersey tea.
Used by Menomini and Ojibwa (Smith, 1923: 49; Gilmore, 1933: 136). Grows from Alabama to Quebec and Manitoba in dry open woods, roadsides, and gravelly shores.

37. *Vitis riparia* Michx. Frost grape, riverbank grape.
Twigs used by Ojibwa (Smith, 1932: 392). Ojibwa also used the sap for medicine (*ibid.*); the Ojibwa, Menomini, Sauk-Fox, and Iroquois used the fruit for food. Grows from Tennessee to Quebec and Manitoba in rich thickets and river banks.

38. *Hypericum virginicum* (L.) Marsh. St. John's wort.
Used by Potawatomi (Smith, 1933: 60). Grows from Florida to Newfoundland and Manitoba in wet sands, bogs, and swamps.

39. *Viola conspersa* Reichenb. American dog violet.
Used by Ojibwa (Smith, 1932: 392). Grows from Maryland to Quebec and Minnesota in meadows, damp woods, bottomlands, etc.

40. *Dirca palustris* L. Leatherwood, *micopy*.
Used by Ojibwa and Potawatomi (Smith, 1932: 390; 1933: 85). Ojibwa, Menomini, Potawatomi, and Winnebago also used it for fiber. Grows from the Gulf of Mexico to northern Lake Superior (frequent throughout Michigan) in damp, rich woods.

41. *Aralia racemosa* L. Spikenard.
Root used by Menomini (Smith, 1923: 24). Menomini and Potawatomi also used it for food; Ojibwa and Potawatomi used it as medicine (Densmore, 1928: 287; Smith, 1933: 41). Grows from Alabama to central Ontario in rich woods and thickets.

42. *Osmorrhiza longistylis* (Torr.) DC. Sweet cicely, anise root.
Root used by Ojibwa and Potawatomi (Smith, 1932: 391; 1933: 86). Grows from Kentucky to Quebec and Alberta in rich, often alluvial woods and thickets.

43. *Chimaphila umbellata* (L.) Bart. Pipsissewa, wintergreen, waxflower.
Used by Ojibwa (Smith, 1932: 368). Grows from Georgia to Quebec and western Ontario in dry woods.

44. *Pyrola rotundifolia* L. Shinleaf.
Dried leaves used by Ojibwa for tea drunk for good luck in hunting (Smith, 1932: 430). Grows from Kentucky to Quebec and Minnesota in woods and clearings.

45. *Vaccinium angustifolium* Ait. Low sweet blueberry, sweet hurts.
Used by Ojibwa (Smith, 1932: 369). Ojibwa, Menomini, and Potawatomi also used the berry as food. Grows from northern Indiana to James Bay in dry, rocky or sandy soil, burns, clearings, dry open barrens, and peats.

46. *Vaccinium oxycoccos* L. Small cranberry.
Used by Ojibwa (Smith, 1932: 369). Ojibwa, Menomini, Potawatomi, and Iroquois also used the berry as food. Grows from Ohio to northern Canada in cold sphagnum bogs.

47. *Gentiana flavida* Gray. Yellowish gentian.
Root used by Potawatomi (Smith, 1393: 58). Grows from Tennessee to southern Ontario, Michigan, and Minnesota in damp woods, prairies, and meadows.

48. *Verbena hastata* L. Blue vervain, simpler's joy.
Used by Menomini and Ojibwa (Smith, 1923: 58; Densmore, 1928: 294). Grows from Florida and Texas to Nova Scotia and British Columbia in swales, damp thickets, and shores.

49. *Monarda fistulosa* L. Wild bergamot.
Used by Menomini (Smith, 1923: 39). Ojibwa and Fox also used the plant as medicine (Smith, 1933: 61). Grows from Alabama to Quebec and Minnesota in dry thickets, clearings, and borders of woods.

50. *Galium aparine* L. Spring cleavers, goosegrass.
Used by Ojibwa (Smith, 1932: 386). Grows from Florida and Texas to Alaska and Newfoundland in rich woods, thickets, seashores, and waste ground; a weedy species. Charred Galium seeds are abundant in the Juntunen site on Bois Blanc Island, Michigan. This species is reported as an element of the modern flora of the island.

51. *Galium tinctorium* L. Small cleaver.
Used by the Ojibwa (Smith, 1932: 386). Grows from Kentucky to Newfoundland and Alaska in swamps and damp places.

52. *Galium triflorum* Michx. Fragrant bedstraw.
Used by Menomini (Smith, 1923: 51). Grows from Mexico to Newfoundland and Alaska in woods and thickets.

53. *Mitchella repens* L. Partridge berry.
Used by Menomini (Smith, 1923: 51). Ojibwa also used it for medicine (Gilmore, 1933: 141) and smoking material. Grows from Florida and Texas to Newfoundland and Minnesota in dry or moist knolls in woods.

54. *Diervilla lonicera* Mill. Bush honeysuckle.
Used by Menomini (Smith, 1923: 27). Ojibwa and Potawatomi for medicine (Smith, 1932: 360; 1933: 45). Grows from Ohio to Newfoundland and Manitoba in dry woods, clearings, and rocky places.

55. *Lonicera canadensis* Bartr. Fly honeysuckle.
Root used by Potawatomi (Smith, 1933: 46). Menomini used the bark for medicine (Smith, 1923: 27). Grows from Ohio to Quebec and Saskatchewan in moist woods.

56. *Viburnum acerifolium* L. Dockmackie, arrowwood.
Used by Menomini and Ojibwa (Smith, 1923: 28; Densmore, 1928: 294). Grows from Tennessee to Quebec and Minnesota in dry or rocky woods.

57. *Sambucus canadensis* L. Common elder.
Used by Menomini and Ojibwa (Smith, 1923: 27; Gilmore, 1933: 142). Ojibwa, Sauk-Fox, and Iroquois also used the berry for food. Grows from the Gulf of Mexico to central Ontario in rich, moist ground, stream borders, fence rows, and following settlements.

58. *Valeriana uliginosa* (T & G) Rydb. Swamp valerian.
Root used by Menomini (Smith, 1923: 57). Menomini fish lure. Grows from Michigan to Quebec and western Ontario in calcareous swamps and wet woods.

59. *Eupatorium perfoliatum* L. Boneset, thoroughwort.
Used by Menomini and Ojibwa (Smith, 1923: 30; Gilmore, 1933: 142). Grows from Florida and Texas to Quebec and Manitoba in low woods or thickets, swales, wet shores, etc., usually common.

60. *Solidago flexicaulis* L. Goldenrod.
Used by Potawatomi and Ojibwa (Smith, 1933: 53; Densmore, 1928: 293). Grows from Tennessee to Quebec and North Dakota in rich woods, thickets, and cool slopes.

61. *Solidago canadensis* L. Goldenrod.
Flowers used by Potawatomi (Smith, 1933: 53). Grows from Tennessee to Quebec and Saskatchewan in moist to dryish thickets, roadsides, clearings, and slopes.

62. *Solidago gigantea* Ait. Goldenrod.
Flowers used by Potawatomi and Menomini (Smith, 1933: 53; 1923: 31). Grows from Florida and Texas to Prince Edward Island and Oregon in damp thickets, alluvial or rich soil, etc.

63. *Solidago graminifolia* (L.) Salisb. Fragrant goldenrod.
Flowers used by Potawatomi and Ojibwa (Smith, 1933: 53; 1932: 366). Also smoked by Ojibwa to attract deer. Grows from Ohio to James Bay in damp to dryish shores, thickets, etc.

64. *Aster shortii* Lindl. Aster, starwort, frostflower.
Flowers used by Potawatomi (Smith, 1933: 50). Grows from Alabama to Pennsylvania and Wisconsin in open woods, thickets, and rocky slopes.

65. *Erigeron philadelphicus* L. Philadelphia fleabane.
Used by Ojibwa (Smith, 1932: 364). Also an Ojibwa smoking material. Grows from Florida and Texas to Newfoundland and British Columbia in rich thickets, alluvium, shores, and springy slopes.

66. *Rudbeckia hirta* L. Blackeyed Susan, cone flower.
Used by Potawatomi and Ojibwa (Smith, 1933: 53; Gilmore, 1933: 143). Potawatomi also used it as a dye. Grows from Alabama to Massachusetts and Michigan in open woods, thickets, barrens, fields, and waste ground.

67. *Helenium autumnale* L. Sneezeweed.
Used by Menomini (Smith, 1923: 30). Grows from Kentucky to Quebec and Minnesota on shores, meadows, and low thickets.

68. *Artemisia canadensis* Michx. Canada wormwood.
Used by Menomini (Smith, 1923: 29). Grows from northern Michigan to Labrador and British Columbia in gravels, sands, and on calcareous rocks and cliffs.

APPENDIX C

OTHER MEDICINAL PLANTS

The following list includes 207 medicinal plants but does not include the medicinal teas. The order of the entries follows *Gray's Manual of Botany* (Fernald, 1950). The entries include scientific name, common name, tribes reported as using the medicine, part of the plant used when known, other uses of the plant, and occasionally other comments.

1. *Calvatia craniiformis* Schw. Giant puffball. Ojibwa (Smith, 1932: 370).
2. *Lycoperdon subincarnatum* Ph. Pinkish puffball. Potawatomi (Smith, 1933: 64); spores.
3. *Lycopodium clavatum* L. Clubmoss. Potawatomi (Smith, 1933: 64); spores.
4. *Lycopodium obscurum* L. Ground pine. Ojibwa and Potawatomi (Smith, 1932: 375; 1933: 64); spores.
5. *Lycopodium complanatum* L. Ground pine. Ojibwa (Smith, 1932: 375).
6. *Botrychium virginianum* (L.) Sw. Rattlesnake fern. Ojibwa and Potawatomi (Smith, 1932: 377; 1933: 67).
7. *Onoclea sensibilis* L. Sensitive fern. Ojibwa (Smith, 1932: 382); root.
8. *Tsuga canadensis* (L.) Carr. Hemlock. Ojibwa and Potawatomi (Smith, 1932: 380; 1933: 71). Also used for beverage and dye.
9. *Picea mariana* (Mill.) BSP black spruce. Ojibwa and Potawatomi (Smith, 1932: 379; 1933: 70). Also utilized technologically.
10. *Pinus resinosa* Ait. Red or Norway pine. Ojibwa and Potawatomi (Smith, 1932: 379; 1933: 70). Also utilized technologically.
11. *Pinus banksiana* Lamb. Jack pine. Ojibwa, Menomini, and Potawatomi (Smith, 1932: 379; 1923: 45; 1933: 70). Also used technologically.

12. *Juniperus communis* L. Common juniper. Ojibwa and Potawatomi (Gilmore, 1933: 124; Smith, 1933: 69). Also utilized technologically.

13. *Juniperus virginiana* L. Red cedar. Ojibwa (Densmore, 1928: 290). Also utilized technologically.

14. *Sagittaria latifolia* Willd. Duck potato, *wapatoo*. Ojibwa and Potawatomi (Densmore, 1928: 292; Smith, 1933: 37). Also used for food.

15. *Glyceria canadensis* (Michx.) Trin. Rattlesnake grass. Ojibwa (Smith, 1932: 371); root.

16. *Hordeum jubatum* L. Potawatomi and Ojibwa (Smith, 1933: 59; Densmore, 1928: 289); root.

17. *Andropogon gerardi* Vitman. Beard grass. Ojibwa (Densmore, 1929: 286). Found in Ohio cave textiles by Whitford.

18. *Arisaema triphyllum* (L.) Schott. Jack-in-the-pulpit. Ojibwa and Menomini (Smith, 1932: 356; Densmore, 1928: 287). Also used for food.

19. *Symplocarpus foetidus* (L.) Nutt. Skunk cabbage. Ojibwa and Menomini (Gilmore, 1933: 124; Smith, 1923: 23). Also used for food.

20. *Calla palustris* L. Wild calla, Water arum. Potawatomi (Smith, 1933: 40); root.

21. *Acorus calamus* L. Sweet flag. Ojibwa, Menomini, and Potawatomi (Smith, 1932: 355; 1923: 22; 1933: 39); root. Also utilized technologically and as a charm.

22. *Uvularia grandiflora* Sm. Large-flowered bellwort. Ojibwa, Menomini, and Potawatomi (Smith, 1932: 374; 1923: 41; 1933: 64); root.

23. *Allium stellatum* Fraser. Wild onion. Ojibwa (Densmore, 1928: 286).

24. *Allium tricoccum* Ait. Wild leek, ramp. Ojibwa (Densmore, 1928: 286). Also used for food.

25. *Lilium philadelphicum* L. Wood lily. Ojibwa (Gilmore, 1933: 125). Also used for food.

26. *Lilium canadense* L. Wild yellow lily. Ojibwa (Densmore, 1928: 290). Also used for food.

27. *Clintonia borealis* (Ait.) Raf. Corn lily, Blue-bead lily. Ojibwa and Potawatomi (Smith, 1932: 373; 1933: 62); root.

28. *Smilacina racemosa* (L.) Dasf. False Solomon's-seal. Ojibwa, Menomini, and Potawatomi (Smith, 1932: 374; 1923: 41; 1933: 63); root. Also used for food.

29. *Maianthemum canadense* Dasf. Wild lily-of-the-valley. Ojibwa and Potawatomi (Smith, 1932: 373; 1933: 62). Also used for food.

30. *Streptopus roseus* Michx. Twisted-stalk, Rose mandarin. Ojibwa and Potawatomi (Smith, 1932: 374; 1933: 63); root.

31. *Polygonatum pubescens* (Willd.) Pursh Solomon's-seal. Ojibwa and Menomini (Smith, 1932: 374; Smith, 1923: 41); root.

32. *Smilax herbacea* L. Carrion flower, Jacob's ladder. Ojibwa (Smith, 1932: 374); root.

33. *Iris versicolor* L. Blue flag, poison flag. Ojibwa and Potawatomi (Smith, 1932: 371; 1933: 60); root. Also used for technology and as a charm.

34. *Cypridpedium calceolus* L. var. *pubescens* (Willd.) Correll. Yellow lady's-slipper. Ojibwa and Menomini (Smith, 1932: 377; 1923: 40). Also used for a charm.

35. *Cypripedium reginae* Walt. Showy lady's-slipper. Ojibwa (Densmore, 1928: 289).

36. *Cypripedium acaule* Ait. Stemless lady's-slipper. Menomini (Smith, 1923: 44); root.

37. *Goodyera repens* (L.) R. Br. var. *ephoides* Fern. Rattlesnake plantain. Potawatomi (Smith, 1933: 67).

38. *Malaxis unifolia* Michx. Green adder's-mouth. Ojibwa (Smith, 1932: 377); root.

39. *Salix humilis* Marsh. Small pussy willow. Menomini and Potawatomi (Smith, 1933: 81); root.

40. *Salix lucida* Muhl. Shining willow. Ojibwa (Smith, 1932: 388). Also used for smoking material.

41. *Salix pedicellaris* Pursh. Bog willow. Ojibwa (Smith, 1932: 388); bark.

42. *Salix gracilis* Anderss. Willow. Potawatomi (Smith, 1933: 82).

43. *Populus balsamifera* L. Balsam poplar, hackmatack, tacamahac. Ojibwa, Menomini, and Potawatomi (Smith, 1932: 387; 1923: 52; 1933: 80); resin.

44. *Populus tremuloides* Michx. Quaking aspen. Ojibwa (Smith, 1932: 388); bark. Also used for food.

45. *Comptonia peregrina* (L.) Coult. Sweet fern. Ojibwa and Potawatomi (Smith, 1932: 375; 1933: 65). Also used for beverage, technology, and ceremonial incense.

46. *Juglans cinerea* L. Butternut. Potawatomi and Menomini (Smith, 1933: 60; 1923: 68); sap. Also used for food and dye.

47. *Corylus americana* Walt. Hazlenut. Ojibwa (Smith, 1932: 359). Also used for food, dye, and technology.

48. *Corylus cornuta* Marsh. Beaked hazelnut. Ojibwa and Potawatomi (Smith, 1932: 359; 1933: 44). Also used for food and technology.

49. *Ostrya virginiana* (Mill.) K. Koch. American hop-hornbeam, ironwood. Potawatomi and Ojibwa (Smith, 1933: 44; Densmore, 1928: 291); bark. Also utilized technologically.

50. *Betula lenta* L. Sweet birch, black birch. Ojibwa (Gilmore, 1933: 128); bark.

51. *Betula pumila* L. Low birch, swamp birch. Ojibwa (Smith, 1932: 359). Also utilized technologically.

52. *Alnus rugosa* (Du Roi) Spreng. Speckled alder. Ojibwa, Menomini, and Potawatomi (Smith, 1932: 358; 1923: 26; 1933: 43); bark, root. Also used for dye.

53. *Fagus grandifolia* Ehrh. Beech. Ojibwa, Menomini, and Potawatomi (Gilmore, 1933: 128; Smith, 1923: 36; 1933: 58) leaves. Also used for food.

54. *Quercus alba* L. White oak. Menomini (Smith, 1923: 36); cambium. Also used for food.

55. *Quercus macrocarpa* Michx. Mossy cup oak, burr oak. Ojibwa (Smith, 1932: 369). Also used for food and dye.

56. *Quercus rubra* L. Red oak. Ojibwa and Potawatomi (Smith, 1932: 369; 1933: 58); cambium. Also used for food and dye.

57. *Quercus velutina* Lam. Black oak. Menomini (Smith, 1923: 36); bark. Also used for food and dye.

58. *Ulmus americana* L. White elm. Potawatomi (Smith, 1933: 86); bark. Also utilized technologically.

59. *Asarum canadense* L. Wild ginger. Ojibwa and Menomini (Smith, 1932: 357; 1923: 24); root. Also used for food and flavoring.

60. *Polygonum careyi* Olney. Persicaria. Potawatomi (Smith, 1933: 72).

61. *Polygonum punctatum* Ell. Water smartweed. Ojibwa (Densmore, 1928: 291).

62. *Chenopodium capitatum* (L.) Aschers. Strawberry blight, Indian paint. Potawatomi (Smith, 1933: 47). Also used for dream dance paint.

63. *Myrabilis nyctaginea* (Michx.) MacM. Four-o'clock. Ojibwa (Smith, 1932: 375); root.

64. *Nuphar advena* (Ait.) Ait.f. Yellow pond lily. Ojibwa, Menomini, and Potawatomi (Smith, 1932: 376; 1923: 43; 1933: 65); root.

65. *Nymphaea odorata* Ait. Pond lily. Ojibwa and Potawatomi (Smith, 1932: 376; 1933: 65); root. Also used for food.

66. *Ranunculus pensylvanicus* L. F. Bristly crowfoot. Potawatomi (Smith, 1933: 75). Also used for dye and smoked to attract deer.

67. *Thalictrum dasycarpum* Fisch. and Call. Purple meadowrue. Ojibwa and Potawatomi (Smith, 1932: 383; 1933: 75); root.

68. *Hepatica acutiloba* DC. Liverleaf. Menomini (Smith, 1923: 48); root.

69. *Anemone virginiana* L. Thimbleweed. Menomini (Smith, 1923: 48); root.

70. *Anemone canadensis* L. Canada anemone. Ojibwa (Smith, 1932: 382); root.

71. *Anemone patens* L. Pasque flower. Ojibwa (Densmore, 1928: 292).

72. *Caltha palustris* L. Marsh marigold. Ojibwa (Gilmore, 1933: 130); root. Also used for food.

73. *Coptis groenlandica* (Oeder.) Fern. Gold thread, cankerroot. Ojibwa, Menomini, and Potawatomi (Smith, 1932: 383; 1923: 48; 1933: 74); root. Also used for dye.

74. *Aquilegia canadensis* L. Wild columbine. Ojibwa (Smith, 1932: 383); root.

75. *Actaea pachypoda* Ell. White baneberry, white cohosh. Ojibwa (Gilmore, 1933: 130).

76. *Caulophyllum thalictroides* (L.) Michx. Blue cohosh, papoose root. Ojibwa, Menomini, and Potawatomi (Smith, 1932: 358; 1923: 25; 1933: 43); root.

77. *Menispermum canadense* L. Canada moonseed. Ojibwa (Smith, 1932: 375); root.

78. *Lindera benzoin* (L.) Blume. Spice bush, Benjamin bush. Ojibwa (Gilmore, 1933: 131). Also used for flavoring.

79. *Sassafras albidum* (Nutt.) Nees. Ojibwa (Gilmore, 1933: 124). Also used for beverage and for flavoring.

80. *Sanguinaria canadensis* L. Bloodroot. Ojibwa, Menomini, and Potawatomi (Smith, 1932: 377; 1923: 78; 1933: 68); root. Also used for dye and face paint.

81. *Corydalis aurea* Willd. Golden corydalis. Ojibwa (Smith, 1932: 370); root.

82. *Lepidium virginicum* L. Poor-man's pepper. Menomini (Smith, 1923: 33).

83. *Sarracenia purpurea* L. Pitcher plant. Ojibwa, Menomini, and Potawatomi (Smith, 1932: 389; 1923: 52; 1933: 82); root.

84. *Saxifraga pensylvanica* L. Swamp saxifrage, wild beet. Menomini (Smitn, 1923: 53).

85. *Ribes cynosbati* L. Prickly gooseberry, dogberry. Potawatomi (Smith, 1933: 82). Also used for food.

86. *Ribes oxyacanthoides* L. Smooth gooseberry. Ojibwa (Densmore, 1928: 289). Also used for food.

87. *Ribes triste* Pall. Red currant. Ojibwa (Smith, 1932: 389). Also used for food.

88. *Hamamelis virginiana* L. Witch hazel. Ojibwa, Menomini, and Potawatomi (Gilmore, 1933: 131; Smith, 1923: 37; Smith, 1933: 59); twigs. Also used for a beverage and seeds used as sacred beads in medicine ceremony.

89. *Pyrus americana* (Marsh.) DC. Mountain ash. Ojibwa and Potawatomi (Reagan, 1928: 236; Smith, 1933: 78); bark. Also utilized technologically.

90. *Amelanchier canadensis* (L.) Medic. Juneberry, service berry. Ojibwa (Densmore, 1928: 286). Also used for food.

91. *Amelanchier laevis* Wieg. Juneberry, service berry. Ojibwa (Smith, 1932: 384); bark. Also used for food.

92. *Crataegus* sp. Ojibwa; root, fruit, and bark (Gilmore, 1933: 132; Smith, 1932: 384). Also used for food, technology, and smoked to attract deer.

93. *Crataegus chrysocarpa* Ashe. Potawatomi (Smith, 1933: 76); fruit. Also used for food.

94. *Fragaria virginiana* Duchesne. Wild strawberry. Ojibwa (Smith, 1932: 384); root. Also used for food.

95. *Fragaria vesca* Porter. Sow-teat strawberry. Potawatomi (Smith, 1933: 76); root. Also used for food.

96. *Potentilla arguta* Pursh. Cinquefoil. Ojibwa (Densmore, 1928: 289).

97. *Potentilla palustris* (L.) Scop. Marsh cinquefoil. Ojibwa (Smith, 1932: 384).

98. *Potentilla norvegica* L. Rough cinquefoil. Ojibwa and Potawatomi (Smith, 1932: 384; 1933: 77).

99. *Geum macrophyllum* Willd. Large-leaved avens. Ojibwa (Smith, 1932: 384).

100. *Geum triflorum* Pursh. Prairie smoke. Ojibwa (Densmore, 1928: 293).

101. *Rubus strigosus* Michx. Raspberry. Ojibwa, Menomini, and Potawatomi (Smith, 1932: 386; 1923: 50; 1933: 79); root. Also used for food.

102. *Rubus occidentalis* L. Black raspberry. Ojibwa and Menomini (Densmore, 1928: 292; Smith, 1923: 71); root. Also used for food.

103. *Agrimonia gryosepala* Wallr. Cocklebur. Ojibwa and Mascouten (Smith, 1932: 383; 1933: 76).

104. *Rosa blanda* Ait. Smooth rose. Ojibwa and Potawatomi (Smith, 1932: 385; 1933: 78); root.

105. *Rosa arkansana* Porter. Rose. Ojibwa (Densmore, 1928: 292).

106. *Prunus americana* Marsh. Wild plum. Ojibwa (Densmore, 1928: 291). Also used for food.

107. *Psoralea argophylla* Pursh. Scurvey pea. Ojibwa (Densmore, 1928: 292).

108. *Petalostemum purpureum* (Vent.) Rydb. Prairie clover. Ojibwa (Densmore, 1928: 291).

109. *Astragalus caryocarpus* Ker. Ground plum, buffalo bean. Ojibwa (Densmore, 1928: 287).

110. *Lathyrus venosus* Muhl. Vetchling, wild pea. Ojibwa (Densmore, 1928: 290). Also used for food and as a charm for success.

111. *Lathyrus ochroleucus* Hook. Creamy vetchling. Ojibwa (Smith, 1932: 372). Also used for food and as a charm for success.

112. *Amphicarpa bracteata* (L.) Fern. Hog .peanut, ground bean. Ojibwa (Densmore, 1928: 289). Also used for food.

113. *Geranium maculatum* L. Spotted cranebill. Ojibwa and Menomini (Smith, 1932: 370; 1923: 36); root.

114. *Ptelia trifoliata* L. Hop tree, wafer ash. Sacred Menomini medicine (Smith, 1923: 51).

115. *Euphorbia corollata* L. Flowering spurge. Ojibwa (Smith, 1932: 369); root.

116. *Rhus typhina* L. Staghorn sumac. Ojibwa, Menomini, and Potawatomi (Smith, 1932: 354; 1923: 22; 1933: 38); root bark, cambium, and berries. Also used for food, beverage, dye, and smoking material.

117. *Rhus glabra* L. Smooth sumac. Ojibwa (Gilmore, 1933: 135). Also used for beverage, dye, and smoking material.

118. *Ilex verticillata* (L.) Gray. Winterberry, black alder. Ojibwa (Smith, 1932: 355).

119. *Celastrus scandens* L. Climbing bittersweet, waxwork, staff tree. Ojibwa (Smith, 1932: 362). Also used for food.

120. *Acer spicatum* Lam. Mountain maple. Ojibwa and Potawatomi (Smith, 1932: 353; 1933: 37); cambium.

121. *Acer rubrum* L. Red maple. Ojibwa and Potawatomi (Smith, 1932: 353; 1933: 37); cambium.

122. *Acer negundo* L. Box elder. Ojibwa (Smith, 1932: 353). Also used for food.

123. *Impatiens pallida* Nutt. Pale jewelweed. Ojibwa and Potawatomi (Smith, 1932: 357; 1933: 42). Also used for dye.

124. *Rhamnus alnifolia* L'Her. Alder buckthorn, dwarf alder. Potawatomi (Smith, 1933: 75); bark.

125. *Ceanothus ovatus* Desf. Inland jersey tea, smaller red-root. Ojibwa (Densmore, 1928: 288).

126. *Vitis* sp. Grape. Ojibwa (Gilmore, 1933: 136). Also used for food.

127. *Hypericum pyramidatum* Ait. Great St. John's-wort. Menomini (Smith, 1923: 37); root.

128. *Viola pubescens* Ait. Downy yellow violet. Potawatomi (Smith, 1933: 87); root.

129. *Epilobium angustifolium* L. Fireweed, great willow-herb. Ojibwa, Menomini, and Potawatomi (Smith, 1932: 376; 1923: 43; 1933: 66); root.

130. *Oenothera biennis* L. Evening primrose. Ojibwa and Potawatomi (Smith, 1932: 376; 1933: 66).

131. *Panax quinquefolium* L. Ginseng, sang. Potawatomi and Menomini (Smith, 1933: 41; 1923: 24). Also sold to whites.

132. *Aralia nudicaulis* L. Wild sarsaparilla. Ojibwa and Potawatomi (Smith, 1932: 356; Smith, 1933: 40); root. Also used for a charm.

133. *Taenidia integerrima* (L.) Drude. Yellow pimpernel. Menomini (Smith, 1923: 56). Also smoked as a hunting charm.

134. *Sanicula marilandica* L. Sanicle, black snakeroot. Ojibwa (Smith, 1932: 391); root.

135. *Sanicula canadensis* L. Sanicle, black snakeroot. Ojibwa (Densmore, 1928: 293).

136. *Osmorrhiza claytoni* (Michx.) C.B. Clarke. Sweet jarvil, Wooly sweet cicely. Menomini and Ojibwa (Smith, 1923: 72; Densmore, 1928: 290).

137. *Thaspium barbinode* (Michx.) Nutt. Meadow parsnip. Ojibwa (Densmore, 1928: 293).

138. *Angelica atropurpurea* L. Angelica. Menomini (Smith, 1923: 55); root.

139. *Heracleum maximum* Bartr. Cow parsnip, masterwort. Ojibwa (Smith, 1932: 390); root. Also used for food, hunting and fishing charm, and evil medicine used by sorcerers.

140. *Cornus alternifolia* L. Pagoda dogwood, Green osier. Ojibwa, Menomini, and Potawatomi (Smith, 1932: 366; 1923: 32; 1933: 54); bark. Also used for smoking material.

141. *Cornus stolonifera* Michx. Red osier. Ojibwa and Potawatomi (Densmore, 1928: 288; Smith, 1933: 55); root, bark. Also used for technology, dye, and smoking material.

142. *Cornus canadensis* L. Bunchberry, dwarf cornel. Ojibwa (Smith, 1932: 366). Also used for food.

143. *Cornus florida* L. Flowering dogwood. Ojibwa (Gilmore, 1933: 138).

144. *Cornus amomum* Mill. Red willow, silky cornel, kinnikinnik. Menomini (Smith, 1923: 32). Also used for smoking material.

145. *Pyrola elliptica* Nutt. Wintergreen, shinleaf. Ojibwa (Gilmore, 1933: 138).

146. *Ledum groenlandicum* Oeder. Labrador tea. Ojibwa and Potawatomi (Densmore, 1928: 290; Smith, 1933: 57); leaves. Also used for food, beverage, and dye.

147. *Kalmia angustifolia* L. Sheep laurel. Montagnais (Tantaquidgear 1932: 266).

148. *Chamaedaphne calyculata* (L.) Moench. Potawatomi (Smith, 1933: 56); leaves. Beverage.

149. *Arctostaphylos uva-ursi* (L.) Spreng. Bearberry. Ojibwa (Densmore, 1928: 287). Also used for food, utility, as a charm, and for smoking materials.

150. *Gaultheria procumbens* L. Teaberry, wintergreen. Ojibwa, Menomini, and Potawatomi (Smith, 1932: 369; 1923: 35; 1933: 56); leaf tea (source of methyl salicylate). Also used for food, beverage, and flavoring.

151. *Vaccinium myrtilloides* Michx. Sour-top blueberry. Potawatomi (Smith, 1933: 57); root. Also used for food.

152. *Fraxinus nigra* Marsh. Black ash. Ojibwa and Menomini (Densmore, 1929: 162). Also used for technology and dye.

153. *Menyanthes trifoliata* L. Buckbean. Menomini (Smith, 1923: 36).

154. *Apocynum androsaemifolium* L. Spreading dogbane. Ojibwa and Potawatomi (Smith, 1932: 354; 1933: 38); root. Also utilized technologically and eaten ceremonially.

155. *Asclepias tuberosa* L. Butterfly weed, pleurisy root. Menomini—one of their most important medicines (Smith, 1923: 25). Also used for food and technology.

156. *Asclepias incarnata* L. Swamp milkweed. Ojibwa (Densmore, 1928: 287). Also used for food and technology.

157. *Asclepias syriaca* L. Common milkweed. Ojibwa and Potawatomi (Smith, 1932: 357; 1933: 42). Also used for food and technology, and as a charm.

158. *Hydrophyllum virginianum* L. John's cabbage, Virginia waterleaf. Ojibwa and Menomini (Smith, 1932: 371; 1923: 68); root. Also used for food.

159. *Cynoglossum boreale* Fern. Northern wild comfrey. Ojibwa (Smith, 1932: 359).

160. *Scutellaria epilobiifolia* A. Hamilton. Marsh skullcap. Ojibwa (Smith, 1932: 372).

161. *Agastache foeniculum* (Pursh.) Ktze. Blue giant hyssop. Ojibwa (Densmore, 1928: 286).

162. *Stachys palustris* L. Woundwort. Ojibwa (Densmore, 1928: 293).

163. *Pycnanthemum virginianum* (L.) Durand and Jackson. Ojibwa (Densmore, 1928: 290). Also used for food and flavoring.

164. *Mentha arvensis* L. Mint. Ojibwa, Menomini, and Potawatomi (Smith, 1932: 371; 1923: 39; 1933: 61); leaf tea. Also used for beverage and flavoring.

165. *Mimulus glabratus* HBK. Monkey flower. Potawatomi (Smith, 1933: 83).

166. *Veronicastrum virginicum* (L.) Farus. Culver's root. Ojibwa and Menomini (Densmore, 1928: 294; Smith, 1923: 53); root.

167. *Castilleja coccinea* (L.) Spreng. Scarlet painted cup. Ojibwa (Densmore, 1928: 288).

168. *Melampyrum lineare* Lam. Cow wheat. Ojibwa (Smith, 1932: 389).

169. *Pedicularis canadensis* L. Common lousewort, Wood betony. Ojibwa (Gilmore, 1933: 140). Also used for an aphrodisiac and a love charm.

170. *Phryma leptostachya* L. Lopseed. Ojibwa (Densmore, 1928: 291).

171. *Plantago rugelii* Dcne. Plantain. Menomini (Smith, 1923: 46); leaves.

172. *Galium trifidum* L. Small bedstraw. Ojibwa (Smith, 1932: 387).

173. *Lonicera dioica* L. Honeysuckle. Ojibwa (Gilmore, 1933: 141).

174. *Linnaea borealis* L. Twin flower. Potawatomi (Smith, 1933: 45).

175. *Symphoricarpus albus* Blake. Snowberry. Ojibwa (Smith, 1932: 361); root.

176. *Viburnum lentago* L. Sheepberry, nannyberry. Ojibwa (Smith, 1932: 361); bark. Berry also used for food.

177. *Viburnum trilobum* Marsh. High-bush cranberry, pimbina. Ojibwa (Smith, 1932: 361); bark. Also used for food and as bait in snares for snowshoe rabbits.

178. *Sambucus pubens* Michx. Red-berried elder. Ojibwa, Menomini and Potawatomi (Smith, 1932: 360; 1923: 27; 1933: 46); bark. Berry also used for food.

179. *Valeriana ciliata* T. & G. Valerian. Menomini (Smith, 1923: 57); root.

180. *Echinocystis lobata* (Michx.) T. & G. Wild balsam apple. Ojibwa and Menomini (Smith, 1932: 367; 1923: 33); root.

181. *Campanula rotundifolia* L. Blue bell, hare bell. Ojibwa (Smith, 1932: 360); root.

182. *Eupatorium maculatum* L. Joe-pye weed. Ojibwa (Densmore, 1928: 289).

183. *Eupatorium purpureum* L. Joe-pye weed. Ojibwa, Potawatomi, and Menomini (Gilmore, 1933: 142; Smith, 1923: 30; Smith, 1933: 52).

184. *Solidago speciosa* Nutt. Goldenrod. Ojibwa (Densmore, 1928: 293).

185. *Solidago juncea* Ait. Goldenrod. Ojibwa (Densmore, 1928: 293).

186. *Solidago uliginosa* Nutt. Goldenrod. Potawatomi (Smith, 1933: 53); root.

187. *Solidago altissima* L. Goldenrod. Ojibwa (Densmore, 1928: 293).

188. *Solidago rigida* L. Goldenrod. Ojibwa (Densmore, 1928: 293).

189. *Aster furcatus* Burgess. Starwort, frostflower. Potawatomi (Smith, 1933: 49).

190. *Aster macrophyllus* L. Starwort, frostflower. Ojibwa (Smith, 1932: 363); root. Also used for food and smoked to attract deer.

191. *Aster nemoralis* Ait. Starwort, frostflower. Ojibwa (Densmore, 1928: 287).

192. *Erigeron canadensis* L. Horseweed, logweed, butterweed. Ojibwa (Densmore, 1928: 289). Also used for smoking material.

193. *Erigeron strigosus* Muhl. Daisy fleabane, whitetop. Ojibwa (Smith, 1932: 364).

194. *Antennaria neodioica* Greene. Lesser cat's foot. Ojibwa (Smith, 1932: 363).

195. *Anaphalis margaritacea* (L.) C.B. Clarke. Pearly everlasting. Ojibwa (Smith, 1932: 362).

196. *Silphium perfoliatum* L. Indian cup. Ojibwa (Smith, 1932: 365).

197. *Heliopsis helianthoides* (L.) Sweet. Oxeye. Ojibwa (Densmore, 1928: 289).

198. *Rudbeckia laciniata* L. Coneflower. Ojibwa (Densmore, 1928: 292).

199. *Achillea lanulosa* Nutt. Wooly yarrow. Ojibwa (Smith, 1932: 362). Also used for ceremonial smoking material.

200. *Artemisia glauca* Pall. Sage. Ojibwa (Densmore, 1928: 287).

201. *Artemisia ludoviciana* Nutt. White sage. Ojibwa (Densmore, 1928: 287). Also used as a charm.

202. *Artemisia frigida* Willd. Prairie sagewort. Ojibwa and Potawatomi. Not native to Potawatomi but planted by them (Densmore, 1928; Smith, 1933: 49).

203. *Petasites palmatus* (Ait.) Gray. Sweet coltsfood. Menomini (Smith, 1923: 31).

204. *Cirsium discolor* (Muhl.) Spreng. Field thistle. Fox (Smith, 1928: 213).

205. *Lactuca canadensis* L. Lettuce. Ojibwa and Menomini (Densmore, 1928: 290; Smith, 1923: 31).

206. *Lactuca biennis* (Moench) Fern. Tall blue lettuce. Ojibwa and Potawatomi (Smith, 1932: 364; 1933: 52). Also used as a hunting lure.

207. *Prenanthes alba* L. White lettuce. Ojibwa (Smith, 1932: 365).

APPENDIX D

CHARM AND CEREMONIAL PLANTS

In the following list of thirty-one plants, the entries include the scientific and common names of each plant, use as a charm, etc., the tribes who used it, and other uses of the plant. The order of entries is taxonomic, following *Gray's Manual of Botany* (Fernald, 1950). Those plants which were smoked essentially as charms are not included in this list but may be found in the list of plants used for smoking.

1. *Carex plantaginea* Lam. Plantain-leaved sedge. Used as a charm against rattlesnakes by Menomini (Smith, 1923: 34).

2. *Acorus calamus* L. Sweet flag. Used as fishing medicine on net by Ojibwa (Smith, 1932: 428). Used for medicine and wigwam thatch.

3. *Arisaema dracontium* (L.) Schott. Dragon root. Used in sacred bundles to give power of supernatural dreams by the Menomini (Smith, 1923: 79).

4. *Uvularia sessifolia* L. Wild oats, bellwort. Used as part of hunting medicine by Ojibwa (Smith, 1932: 430). (Smoked?).

5. *Polygonatum canaliculatum* (Muhl.) Pursh. Solomon's seal. Used as an incense medical charm by Ojibwa (Gilmore, 1933: 125). Root was used as food by the Iroquois.

6. *Smilax tamnoides* L. Chinaroot, hellfetter. Used for malicious magic by Ojibwa (Gilmore, 1933: 126).

7. *Sisyrinchium albidum* Raf. Blue-eyed grass. Used to ward off snakes by the Menomini (Smith, 1923: 38).

8. *Iris versicolor* L. Blue flag, poison flag. Used as a charm against snakes by Ojibwa (Smith, 1932: 430). Used for mats and baskets and also medicine.

9. *Habenaria viridis* (L.) R. Br. Rein orchis, bracted green orchis. Used as a love charm and aphrodisiac by Ojibwa (Smith, 1932: 377, 431).

10. *Cypripedium calceolus* L. var. *pubescens* (Willd.) Correll. Yellow lady's-slipper. Found in Menomini sacred bundles (Smith, 1923: 40). Also used for medicine.

11. *Spiranthes gracilis* (Bigel) Beck. Slender ladies-tresses. Ingredient of hunting charm to bring game to Ojibwa (Smith, 1932: 431). (Smoked?).

12. *Comptonia peregrina* (L.) Coult. Sweet fern. Used for ceremonial incense by Ojibwa (Gilmore, 1933: 127). Also used for beverage, technology, and medicine.

13. *Hepatica americana* (DC) Ker. Liverleaf. Used for charm by Ojibwa. Also used for dye and medicine (Densmore, 1928: 289).

14. *Hamamelis virginiana* L. Witch hazel. Seeds used as the sacred bead in the medicine ceremony by Menomini (Smith, 1923: 37). Also used as beverage and medicine.

15. *Spiraea alba* Du Roi. Meadowsweet. Used as a trapping medicine by Ojibwa (Smith, 1932: 386).

16. *Lathyrus venosus* Muhl. Vetchling, wild pea. Used as charm for success by Ojibwa (Densmore, 1928: 290). Also used for food and medicine.

17. *Polygala senega* L. Seneca snakeroot. Used as charm by Ojibwa (Densmore, 1928: 291). Also used for tonic (*ibid.*). Large quantities pressed for transport to larger towns by Menomini (Hoffman, 1896: 292).

18. *Aralia nudicaulis* L. Wild sarsaparilla. Used as a fish lure on nets by Ojibwa (Smith, 1932: 428). Also used for medicine.

19. *Sium suave* Walt. Water parsnip. Seeds burned for a hunting charm by Ojibwa (Smith, 1932: 432). Also used for food by Chipewyan and Cree.

20. *Heracleum maximum* Bartr. Cow parsnip, masterwort. Root or seeds burned for hunting charm and root used for fishing charm by Ojibwa; Menomini sorcerer's evil medicine (Smith, 1932: 432; 1923: 55). Also used for food and medicine.

21. *Pyrola rotundifolium* L. Shinleaf. Leaves used for tea for hunting success by Ojibwa (1932: 430).

22. *Arctostaphylos uva-ursi* (L.) Spreng. Bearberry. Used for charm by Ojibwa (Densmore, 1928: 287). Used for food, technology, smoking, and medicine.

23. *Asclepias syriaca* L. Common milkweed, silkweed. Used as a hunting charm by Ojibwa (Smith, 1932: 428).

24. *Pedicularis canadensis* L. Common lousewort, wood betony. Root used as a love charm and aphrodisiac by Ojibwa (Smith, 1932: 432). Also used for medicine and for food by Iroquois.

25. *Valeriana uliginosa* (T. & G.) Rydb. Root juice used on hooks to lure fish by Menomini (Smith, 1923: 57). Also used for medicine.

26. *Aster puniceus* L. Aster. Used as a charm by Ojibwa (Densmore, 1928: 287).

27. *Aster umbellatus* Mill. Aster. Flowers smudged to expel spirits from sick by Potawatomi (Smith, 1933: 50).

28. *Gnaphalium obtusifolium* L. Mouse-ear everlasting. Leaves used as a sorcerer's medicine by Menomini (Smith, 1923: 30).

29. *Artemisia ludoviciana* Nutt. White sage. Used as a charm by Ojibwa (Densmore, 1928: 287). Used for medicine.

30. *Lactuca biennis* (Moench.) Fern. Tall blue lettuce. Used as a hunting lure by Ojibwa (Smith, 1932: 429). Used for medicine.

31. *Hieracium canadense* Michx. Canada hawkweed. Flowers used as a hunting lure by Ojibwa (Smith, 1932: 429).

APPENDIX E

PLANTS USED AS SMOKING MATERIALS

In the following list of twenty-seven smoking plants the entries include the scientific and common name of each plant, the plant part smoked, tribes using the plant, and other uses. The order of entries is taxonomic, following *Gray's Manual of Botany* (Fernald, 1950).

1. *Salix lucida* Muhl. Shining willow. Bark included in kinnikinnik smoking mixture by Ojibwa (Smith, 1932: 388). Also used for medicine.

2. *Polygonum coccineum* Muhl. Swamp persicaria. Dried flowers included in hunting medicine smoked in pipes to attract buck deer by Ojibwa (Smith, 1932: 431). Also used for medicinal tea.

3. *Ranunculus pensylvanicus* L.f. Bristly crowfoot. Seeds included in hunting medicine smoked to attract deer by Ojibwa (Smith, 1932: 431). Also used for dye.

4. *Spiraea alba* Du Roi. Meadowsweet. Chippewa name means "pipe-stem wood" (Gilmore, 1933: 132). Root also used as trapping medicine.

5. *Crataegus* sp. Hawthorn, thornapple. Bark included in hunting medicine smoked to attract deer by Ojibwa (Smith 1932: 431). Also used for food, technology, and medicine.

6. *Rhus typhina* L. Staghorn sumac. Leaves smoked with tobacco by Potawatomi (Smith, 1933: 116). Also used for food, beverage, dye, and medicine.

7. *Rhus glabra* L. Smooth sumac. Leaves smoked by Menomini and Ojibwa (Hoffman, 1896: 249; Gilmore, 1933: 135). Also used for beverage, dye, and medicine.

8. *Rhus aromatica* Ait. Fragrant sumac. Leaves smoked by Menomini (Hoffman, 1896: 249).

9. *Cicuta maculata* L. Musquash root. Root smoked to attract deer by Ojibwa (Smith, 1932: 432).

10. *Taenidia integerrima* (L.) Drude. Yellow pimpernel. Seeds smoked in a pipe for luck when hunting by Ojibwa (Smith, 1932: 432). Also used for medicinal tea.

11. *Cornus stolonifera* Michx. Red osier. Bark smoked by Ojibwa, Menomini, and Potawatomi (Jones, 1861: 135; Hoffman, 1896: 250; Smith, 1933: 118). Also used for dye and medicine.

12. *Cornus amomum* Mill. Red willow, silky cornel, kinnikinnik. Bark smoked by Menomini (Smith, 1923: 80).

13. *Cornus rugosa* Lam. Round-leaved dogwood. Bark smoked by Menomini (Densmore, 1929: 145).

14. *Cornus racemosa* Lam. Panicled dogwood. Bark smoked by Ojibwa (Smith, 1932: 399). Used for beverage.

15. *Cornus alternifolia* Lif. Alternate-leaved dogwood. Bark smoked by Ojibwa and Menomini (Smith, 1932: 417; 1923: 32). Also used for medicine.

16. *Arctostaphylos uva-ursi* (L.) Spreng. Bearberry, kinnikinnik. Leaves smoked by Ojibwa, Menomini, and Potawatomi (Reagan, 1928: 238; Hoffman, 1896: 250; Smith, 1933: 118). Also used for food, technology, charm, and medicine.

17. *Trientalis borealis* Raf. Starflower. Root included in hunting medicine smoked to attract deer by Ojibwa (Smith, 1932: 431).

18. *Mitchella repens* L. Partridge berry. Much used for smoking by Bois Fort Chippewa (Reagan, 1928: 239). Also used for food and medicine.

19. *Viburnum dentatum* L. Southern arrowwood. Bark smoked by Ojibwa (Smith, 1932: 417).

20. *Solidago graminifolia* (L.) Salisb. Fragrant goldenrod. Flowers included a hunting medicine smoked to attract deer by Ojibwa (Smith, 1932: 429). Also used for medicine.

21. *Aster macrophyllus* L. Aster, starwort, frostflower. Smoked to attract deer by Ojibwa (Smith, 1932: 429). Also used for food and medicine.

22. *Aster cordifolius* L. Blue wood aster. Root included in hunting medicine smoked to attract deer by Ojibwa (Smith, 1932: 428).

23. *Aster novae-angliae* L. New England aster. Powdered root smoked in earliest times by Ojibwa (Densmore, 1928: 287).

24. *Erigeron philadelphicus* L. Philadelphia fleabane. Flowers included in hunting medicine smoked to attract deer by Ojibwa and smoked in tobacco or kinnikinnik mixture (Smith, 1932: 429, 398). Also used for medicine.

25. *Erigeron canadensis* L. Horseweed, hog weed, butterweed. Flowers smoked as a hunting charm by Ojibwa (Smith, 1932: 429).

26. *Anaphalis margaritacea* (L.) C.B. Clarke. Pearly everlasting. Flowers smoked by Potawatomi to drive away evil spirits that cause sickness (Smith, 1933: 48). Also used for medicine.

27. *Achillea lanulosa* Nutt. Wooly yarrow. Flowers smoked ceremonially by Ojibwa (Smith, 1932: 417). Also used for medicine.

APPENDIX F

DYE PLANTS

The following list includes twenty-five plants that were used in dyeing various materials. The entries include scientific and common names, tribes that used the plant for dye, and other uses. The order of entries is taxonomic, following *Gray's Manual of Botany* (Fernald, 1950).

1. *Tsuga canadensis* (L.) Carr. Hemlock. Used by Ojibwa and Menomini (Smith, 1932: 426; 1923: 78). Also used for beverage, wigwam covering, and medicine.

2. *Thuja occidentalis* L. Arborvitae, northern white cedar. Used by Ojibwa (Stowe, 1940: 12). Also used for technology and medicine.

3. *Myrica gale* L. Sweet gale. Used for porcupine quill dye by Ojibwa (Smith, 1932: 425).

4. *Juglans cinerea* L. Butternut. Used by Ojibwa and Menomini (Smith, 1932: 425; 1923: 78). Also used for food and medicine.

5. *Juglans nigra* L. Black Walnut. Used by Ojibwa (Gilmore, 1933: 127). Also used for food.

6. *Corylus americana* Walt. Hazelnut. Used by Ojibwa (Densmore, 1928: 289). Also used for food, technology, and medicine.

7. *Betula papyrifera* Marsh. Paper birch. Used by Ojibwa (Smith, 1932: 425). Also used for technology.

8. *Alnus rugosa* (Du Roi) Spreng. Speckled alder. Used by Ojibwa, Menomini, and Potawatomi (Smith, 1932: 425; 1923: 78; 1933: 116). Also used for technology and medicine.

9. *Quercus macrocarpa* Michx. Bur oak, mossy cup oak. Used by Ojibwa (Smith, 1932: 425). Also used for food and medicine.

10. *Quercus rubra* L. Red oak. Used by Potawatomi (Smith, 1933: 120). Also used for food and medicine.

11. *Quercus velutina* Lam. Black oak. Used by Ojibwa (Smith, 1932: 425). Also used for food.

12. *Ranunculus recurvatus* Poir. Crowfoot, buttercup. Used by Menomini (Smith, 1923: 79).

13. *Ranunculus pensylvanicus* L.f. Bristly crowfoot. Used for dye by Ojibwa and Potawatomi (Smith, 1932: 426; 1933: 123). Also used for medicine and smoked to attract deer.

14. *Hepatica americana* (DC) Ker. Liverleaf. Used by Potawatomi (Smith, 1933: 123). Also used for medicine and a charm.

15. *Coptis groenlandica* (Oeden.) Fern. Gold thread, cankerroot. Used by Ojibwa and Potawatomi (Smith, 1932: 426; 1933: 122). Also used for medicine.

16. *Sanguinaria canadensis* L. Bloodroot. Used for face paint and dye by Ojibwa, Menomini, and Potawatomi (Smith, 1932: 377; 1923: 78; 1933: 68). Also used for medicine.

17. *Prunus nigra* Ait. Canada plum. Used by Ojibwa (Smith, 1932: 426). Also used for food.

18. *Rhus typhina* L. Staghorn sumac. Used by Ojibwa and Menomini (Smith, 1932: 424; 1923: 77). Also used for food, beverage, smoking, and medicine.

19. *Rhus glabra* L. Smooth sumac. Used by Ojibwa (Smith, 1932: 424). Also used for beverage, smoking, and medicine.

20. *Impatiens pallida* Nutt. Pale jewelweed. Used by Ojibwa, Menomini, and Potawatomi (Smith, 1932: 357; 1923: 78; 1933: 116). Also used for medicine.

21. *Cornus stolonifera* Michx. Red osier. Used by Ojibwa (Densmore, 1928: 288). Also used for smoking and medicine.

22. *Ledum groenlandicum* Oeder. Labrador tea. Used by Potawatomi (Smith, 1933: 120). Also used for food, beverage, and medicine.

23. *Fraxinus nigra* Marsh. Black ash. Used by Ojibwa (Gilmore, 1933: 139). Also utilized technologically.

24. *Lithospermum caroliniense* (Walt.) MacM. Puccoon. Used by Ojibwa (Densmore, 1928: 290).

25. *Rudbeckia hirta* L. Blackeyed susan, coneflower. Used by Potawatomi (Smith, 1933: 117). Also used for medicine.

APPENDIX G

PLANTS UTILIZED FOR TECHNOLOGICAL PURPOSES

In the following list of forty-one plants, the scientific and common names of the plants are followed by the plant part used and its use, the tribes listed as using the plant for that purpose, other uses, range, and habitat preferences. Occasionally other comments are injected. The order of entries is taxonomic and follows *Gray's Manual of Botany* (Fernald, 1950).

1. *Abies balsamea* (L.) Mill. Balsam fir.

Pitch used for sealing canoes by Ojibwa (Smith, 1932: 420). Also used for medicine. Grows from central lower Michigan to Hudson Bay in cool, moist, rich soil, low swampy ground, and (to the north) well-drained hillsides.

2. *Tsuga canadensis* (L.) Carr. Hemlock.

Bark used for wigwam covers by Ojibwa (Gilmore, 1933: 123). Used for beverage, medicine, and dye. Grows from southern Michigan to Lake Superior in well-drained uplands.

3. *Larix laricina* (DuRoi) K. Koch. Larch, tamarack.

Roots used to sew canoes and to make bags by Ojibwa (Smith, 1932: 378). Also used for medicine. Grows from northern Ohio to central Hudson Bay in cold, deep swamp and (to the north) drier uplands.

4. *Pinus strobus* L. White pine.

Pitch used for sealing canoes by Potawatomi and Ojibwa (Smith, 1932: 421; 1933: 122). Also used for food. Grows from southern Michigan to James Bay in light, fertile loam and sandy soils of granitic origin.

5. *Pinus banksiana* Lamb. Jack pine.

Roots used for canoe and other coarse sewing by Ojibwa, Potawatomi, and Menomini (Smith, 1932: 421; 1923: 75; 1933: 113). Grows from central Lower Michigan to Hudson Bay in sandy, sterile soil.

6. *Thuja occidentalis* L. Arborvitae, northern white cedar.

Wood used for canoe ribs, toboggans, sturgeon spear handles, roots used to sew canoes, and bark used for rope, twine, nets, bags, and mats (gathered in June and early July) by Ojibwa (Gilmore, 1933: 123; Smith, 1932: 422; Densmore, 1928: 386); used for bags by Menomini (Smith, 1923: 76); used for mats by Ojibwa and Ottawa (Jones, 1948; Douglas, 1941). Also used for dye and medicine. This and paper birch are the two most useful trees for Ojibwa (Smith, 1932: 421). Grows from southern Michigan to James Bay in moist soil in low swamp and river banks.

7. *Juniperus communis* L. Common juniper.

Bark used for weaving mats and house-building by Ojibwa (Reagan, 1928: 245). Also used for medicine. Grows from Ohio to northern Canada in poor rocky soil and pastures.

8. *Juniperus virginiana* L. Red cedar.

Bark used for weaving mats and bags by Ojibwa and Potawatomi (Stowe, 1940: 13; Whitford, 1941: 7). Also used for medicine. Grows from Alabama to central Lower Michigan in loamy soil on sunny slopes, dry rocky hills, and peaty swamps, and by lakes and streams.

9. *Typha latifolia* L. Common cattail.

Leaves used to make mats for thatching wigwams by Ojibwa, Potawatomi, and Menomini; mats made in summer (Smith, 1932: 390; 1923: 77; 1933: 114). Found archaeologically at Kettle Hill and Ash Caves in Ohio. Grows from Mexico to James Bay in marshes or shallow water.

10. *Hierochloë odorata* (L.) Beauv. Sweet grass, vanilla grass, Indian grass.

Long leaves used for sewing and for weaving bags and baskets by Ojibwa, Potawatomi, Menomini, and Ottawa (Smith, 1932: 419; 1933: 120; 1923: 75; Whitford, 1941: 7). Collected in mid-July to September (Jones, 1936*a*: 22). Grows from southern Ohio to northern Canada in meadows, swales, and shores. It is concentrated on abandoned habitation sites on the Walpole Island Reserve in Ontario (Arnold Pilling, personal communication).

11. *Scirpus validus* Vahl. Great bulrush, soft-stem bulrush.

Stems used for weaving large mats and baskets by Ojibwa, Menomini, and Potawatomi (Kinietz and Jones, 1942; Smith, 1923: 74; 1933: 112). Tubers also used for food by Ojibwa. Stems collected in midsummer. Grows from Mexico to Newfoundland and Alaska in brackish or fresh shallow water and marshes. Found archaeologically at Ash and Canter's Caves and South Park Village in Ohio (UMMAEL Repts. 15 and 71).

12. *Scirpus cyperinus* (L.) Kunth. Wool grass.
Stems used for weaving mats and storage bags by Ojibwa (Smith, 1932: 418). Grows from Tennessee to Newfoundland and Minnesota in wet meadows and swamps.

13. *Acorus calamus* L. Sweet flag.
Leaves used for thatching wigwams by Menomini (Smith, 1923: 22). Root used for medicine and also for charm. Grows from the Gulf of Mexico to James Bay in wet places and borders of quiet water.

14. *Juncus dudleyi* Wieg. Dudley's rush.
Stems used for weaving small, fine mats by Ojibwa (Smith, 1932: 419). Grows from Kentucky to James Bay in damp to dryish calcareous or sweet soil.

15. *Juncus effusus* L. Soft rush.
Stems used for weaving bags, pouches, and mats by Ojibwa (Gilmore, 1933: 125). Grows from Florida and Texas to Newfoundland and Minnesota in swales, wet thickets, bogs, damp open ground, etc.

16. *Juncus stygius* L. Bog rush.
Stems used for weaving mats by Ojibwa (Reagan, 1928: 245). Grows from Michigan and New York to Labrador and Saskatchewan.

17. *Iris versicolor* L. Blue flag, poison flag.
Leaves used for weaving mats and baskets by Potawatomi (Smith, 1933: 120). Also used for medicine by Ojibwa and Potawatomi and as a charm by Ojibwa. Grows from nothern Ohio to Labrador and Manitoba in marshes, meadows, ditches, and turfy shores.

18. *Salix nigra* Marsh. Black willow.
Fiber used for making bags, pouches, fish nets, and cord by Ojibwa, Menomini, Ottawa, and Winnebago (Whitford, 1941: 11). Grows from Louisiana to northern Lake Superior on stream banks, lake shores, and rich low woods.

19. *Salix interior* Rowlee. Sandbar willow.

Used for weaving baskets by Ojibwa (Gilmore, 1933: 126). Grows from Texas to Hudson Bay in alluvial soils, often pioneering on bars and beaches.

20. *Comptonia peregrina* (L.) Coult. Sweet fern.

Leaves used by Ojibwa and Menomini (Smith, 1932: 375; 1923: 42) to line blueberry pails to prevent spoilage. Used for beverage, medicine, and, also, ceremonial incense. Grows from Ohio to southern Canada in open sterile woodlands, clearings, and pastures.

21. *Carya ovata* (Mill.) K. Koch. Shagbark hickory.

Wood used for bows, arrows, and general utility by Ojibwa, Menomini, and Potawatomi (Smith, 1932: 419; Skinner, 1921: 321; Smith, 1933: 113). Grows from Florida and Texas to central Lower Michigan in rich woods, bottoms, and slopes.

22. *Betula papyrifera* Marsh. Paper birch.

Bark used for making canoes, buckets, kitchen utensils, house covers, etc. by Ojibwa, Potawatomi, and other tribes (Smith, 1932: 413-414; 1933: 112). Inner bark also used for dye. Bark collected in late June to early July (Densmore, 1928: 386). This and white cedar are the two most sacred trees of the Ojibwa. Birchbark resists decay (Smith, 1932: 414). Birchbark vessels keep contents from decay, even gummy maple syrup for a year (Densmore 1928: 287); thus storage for winter use may be possible. Grows from southern Michigan to Hudson Bay in rich, moist hillsides and stream, lake, and swamp borders. Bark found archaeologically at twenty-seven locations throughout the Juntunen site on Bois Blanc Island, Michigan.

23. *Ulmus rubra* Muhl. Slippery elm, red elm.

Bark used for boxes and baskets by Potawatomi, to cover wigwam sides by Ojibwa and Potawatomi, and for fiber by Menomini (Smith, 1932: 423; 1933: 115; 1923: 77); identified in Ohio Hopewell fabrics by Whitford (1941: 12). Inner bark also used for food by Iroquois and medicine by Ojibwa and Potawatomi. Grows from Alabama to Lake Superior in bottom lands, stream banks, rich moist hillsides, and rocky ridges and slopes.

24. *Ulmus americana* L. White elm.
Bark used for roofing winter houses by Ojibwa (Gilmore, 1933: 129). Also used for medicine by Potawatomi. Bark peeled in July. Grows from the Gulf of Mexico to James Bay in deep, rich, moist loam, bottomlands, and stream banks.

25. *Urtica gracilis* Ait. Nettle
Fiber used for sewing, twine, and weaving bags by Ojibwa, Menomini, Potawatomi, and Sauk-Fox (Smith, 1932: 423; 1923: 77; 1933: 115; Morse, 1822: 126); identified in Ohio Hopewell and rock shelter fabrics by Whitford (1941: 13). Also used for medicine. Grows from Texas to Alaska and Newfoundland in thickets and rich damp soil.

26. *Laportea canadensis* (L.) Wedd. Wood nettle.
Fiber used for sewing, twine, and weaving bags by Ojibwa and Menomini (Smith, 1932: 423; 1923: 77); identified in Ohio Hopewell and rock shelter fabrics by Whitford (1941: 13). Also used for medicine. This fiber is "as important as basswood bark to the industries of the Chippewa woman" (Densmore, 1929: 153). Grows from Alabama to central Ontario in low woods and stream banks.

27. *Boehmeria cylindrica* L. Stingless nettle, false nettle.
Fiber used for bowstring by Ojibwa (Whitford, 1941: 12). Grows from the Gulf of Mexico to Lake Superior in moist or shady ground.

28. *Asimina triloba* (L.) Dunal Paw-paw.
Bark fiber used for weaving bags by Menomini and Potawatomi (Whitford, 1941: 9); identified in Ohio rock shelter fabrics by Whitford (*ibid.*). Fruit also used for food (Aller, 1954: 65). Grows from Florida and Texas to western New York, southern Michigan, and Illinois (common in Grand and Maple River valleys) in rich woods and alluvium.

29. *Pyrus americana* (Marsh) DC. Mountain ash.
Wood used for canoe ribs, snowshoe frames, lacrosse racquets etc. by Ojibwa because it can be bent into any form (Reagan, 1928: 236). Also used for medicine. Grows from northern Illinois and southern Michigan to James Bay in rich, moist soil on river banks, borders of cold swamps, rocky hillsides and mountains.

30. *Amelanchier* spp. Juneberry, serviceberry, shadbush.
Stalks used to make arrows by Ojibwa "in very old times" (Densmore, 1929: 147). Fruit also used for food.

31. *Crataegus* spp. Hawthorn, red haw, thornapple.
Thorns gathered in summer for sewing buckskin with sinew by Ojibwa (Densmore, 1929: 120; Smith, 1932: 422). Products also used for food, medicine, and smoked to attract deer.

32. *Tilia americana* L. Basswood, *bois blanc*.
Bark fiber used for thread, cordage, weaving bags and baskets, tying wigwam framework, sewing mats, fish nets, snowshoes, and ropes by Ojibwa, Potawatomi, Menomini, Sauk-Fox, and Winnebago (Gilmore, 1933: 136; Smith, 1933: 114; 1923: 76; Whitford, 1941: 11); identified in Ohio Hopewell and rock shelter textiles by Whitford (1941: 12); identified in cord in copper beads from a Glacial Kame site in Canada (Ritchie, 1948). Saplings peeled in spring by Menomini (Smith, 1923: 76). "In the Great Lakes region in aboriginal times basswood apparently was the chief source of fibrous material" (Jones, 1937: 13). Sap, young twigs, and buds used for food by Ojibwa. Grows from Tennessee to central Ontario in rich, well-drained, loamy soils.

33. *Dirca palustris* L. Leatherwood, wicopy.
Fiber used for cordage and weaving bags by Ojibwa, Potawatomi, Menomini, and Winnebago (Gilmore, 1933: 137; Smith, 1933: 114; 1923: 76; Whitford, 1941: 11); identified in Ohio Adena and rock shelter textiles by Whitford (1941: 10). Used also for medicinal tea. Grows from the Gulf of Mexico to northern Lake Superior in damp rich woods.

34. *Nyssa sylvatica* Marsh. Black gum, pepperidge.
Wood used for awl handles, mauls, and war clubs by Ojibwa because it will not check or split; called "awl-wood" (Gilmore, 1933: 138). Grows from Florida and Mexico to southern Ontario and northern Lower Michigan (frequent in southern Lower Michigan) in swamp borders and low, wet lands.

35. *Fraxinus americana* L. White ash.
Wood used for fish spears (called "spear timber") and in canoe and snowshoe manufacture by Ojibwa (Gilmore, 1933: 139; Reagan, 1928: 245). Grows from the Gulf of Mexico to Upper Michigan (common throughout Michigan) in rich, moist, loamy soil and any well-drained situation; common along stream beds.

36. *Fraxinus pennsylvanica* Marsh. Red ash.
Wood used to make splint baskets and spoons, and wood of all three of the ashes used to make bows, arrows, snowshoe frames, sled, and cradle boards by Ojibwa (Smith, 1932: 420). Grows from the Gulf of Mexico to northern Lake Superior in wet or moist rich loam, river banks, and swampy lowlands.

37. *Fraxinus nigra* Marsh. Black ash.
Wood used to make splint baskets by Ojibwa and bows and arrows by Menomini (Smith, 1932: 420; 1923: 75). Also used for dye. Grows from southern Ohio to James Bay in deep, cold swamps, low river banks, and any good soil.

38. *Apocynum androsaemifolium* L. Spreading dogbane.
Fiber used for fine sewing by Ojibwa, Menomini, and Potawatomi (fiber makes a strong fine thread) (Smith, 1932: 413; 1923: 73; 1933: 111). Also eaten ceremonially by Ojibwa (Smith, 1932: 428) and used for medicine by Ojibwa and Potawatomi. Grows from Kentucky to Hudson Bay in dry thickets and borders of woods.

39. *Apocynum cannabinum* L. Indian hemp.
Fiber used for best fine cordage by Ojibwa (Gilmore, 1933: 140); identified in a Sauk-Fox bag, Ohio Hopewell fabric, and Adena fabric by Whitford (1941: 9). Grows from the Gulf of Mexico to northern Lake Superior in open woods, thickets, and borders of woods.

40. *Asclepias incarnata* L. Swamp milkweed.
Fiber used for twine by Ojibwa (Gilmore, 1933: 140); identified in Ohio Hopewell and rock shelter textiles by Whitford (1941: 10). Used for food and medicine also. Grows from Florida to Lake Superior in swamps, wet thickets, and shores.

41. *Asclepias syriaca* L. Common milkweed.
Fiber used for sewing thread and fishlines by Menomini and Potawatomi (Smith, 1923: 74; 1933: 111); identified in a Sauk-Fox bag, Kickapoo string, and Ohio rock shelter fish net (Whitford, 1941: 10). Used for food and medicine and also as a charm to attract deer. Grows from Alabama to James Bay in thickets, roadsides, and dry fields.

The following eleven plants utilized technologically are listed separately because they are judged to be of questionable significance for aboriginal subsistence. The entries include scientific and common names, tribes utilizing the plant, the purpose, and plant part used, and other uses. The sequence of entries is taxonomic, following *Gray's Manual of Botany* (Fernald, 1950).

1. *Picea mariana* (Mill.) BSP. Black spruce.
   Ojibwa--sewing and sealing birchbark canoes with roots and pitch (Smith, 1932: 379). Also used for medicine.

2. *Pinus resinosa* Ait. Red pine, Norway pine.
   Ojibwa--sewing and sealing birchbark canoes with roots and pitch (Smith, 1932: 421). Also used for medicine.

3. *Corylus americana* Walt. Hazelnut.
   Ojibwa--basket ribs and brushes of stems (Smith, 1932: 417). Also used for food, dye, and medicine.

4. *Ostrya virginiana* (Mill.) K. Koch. American hop-hornbeam, leverwood, ironwood.
   Ojibwa--wigwam poles (Densmore, 1929: 23). Also used for medicine.

5. *Betula lutea* Michx. Yellow birch.
   Potawatomi, wigwam poles (Smith, 1933: 112). Sap also used for food.

6. *Betula pumila* L. Low birch, swamp birch.
   Ojibwa--basket ribs of stems (Smith, 1932: 417). Also used for medicine.

7. *Alnus rugosa* (Du Roi) Spreng. Speckled alder.
   Ojibwa--sewing birchbark canoes with roots (Smith, 1932: 358). Also used for dye and medicine.

8. *Fagus grandifolia* Ehrh. Beech.
   Potawatomi--wooden bowls (Smith, 1933: 113). Also used for food and medicine.

9. *Acer saccharum* Marsh. Sugar maple.
   Ojibwa--wooden bowls and food-stirring paddles, arrows (Smith, 1932: 413). Sap also used for food.

10. *Cornus alternifolia* L. Blue dogwood, green osier.
    Ojibwa--thatch with twigs (Smith, 1932: 417). Also used for medicine, smoking material.

11. *Arctostaphylos uva-ursi* Spreng. Bearberry.
    Ojibwa, utility (Densmore, 1928: 287). Also used for food, charm, medicine, and smoking materials.

APPENDIX H

CATALOGUE OF ARCHAEOLOGICAL PLANT REMAINS

(An asterisk (*) indicates plant remains identified entirely, or in part, by the author.)

Site	County	State	UMMAEL Rept. No. and Lab.No.	UMMA Cat. Nos.	Plant Remains
Alhart	Monroe	N.Y.	2 13-17	12607- 12609	corn, beans, squash, hickory nut
Baldwin	Fairfield	Ohio	6 68-71	12639- 12642	corn, hickory nut
Walker-Gilmore.	Cass	Nebr.	8 56-63	12627- 12634	corn, squash, chenopod, hickory nut, walnut
Ash Cave	Hocking	Ohio	15A 111-124	. . . .	bulrush, Indian hemp, butternut, walnut, corn, hickory nut, squash, gourd
Kettle Hill Cave.	Fairfield	Ohio	15B 125-144	. . . .	hickory nut, walnut, hazelnut, squash, corn, cattail, Indian hemp
Canter's Caves .	Jackson	Ohio	15C 145-159	. . . .	Indian hemp, bulrush, gourd, corn, grape, butternut
Castle Creek. . .	Broome	N.Y.	28 393	12977	corn
Kettle Hill Cave.	Fairfield	Ohio	31 398-411	13481- 13485	squash, cattail, hickory nut, walnut
Ash Cave.	Hocking	Ohio	33 413-437	13486- 13489	butternut, hickory nut, blackberry canes, grape, Indian hemp, cattail, bulrush, squash, gourd
Harlan Rock Shelter	Harlan	Ky.	42 474-478	13499- 13502	squash
Stephen De Hart	Powell	Ky.	47 508-510	13532- 13533	squash, gourd, chestnut
Crable	Fulton	Ill.	61 611-623	15918- 15930	corn
Kincaid	Pope and Massac	Ill.	67 629-679	14751- 14802	corn, squash, or gourd
South Park	Cuyahoga	Ohio	71 724	14839	bulrush
Fisher	Will	Ill.	75 1066-1068	15352- 15354	corn, acorn
Sackett.	Ontario	N.Y.	80 1134-1135	15953- 15954	corn
Younge.	Lapeer	Mich.	84 1197, 1209	. . . .	corn
Larsen.	Platte	Nebr.	89 1261-1263	15971- 15973	corn, beans, sunflower
Lynch	Boyd	Nebr.	91 1291-1312	. . . .	corn, beans, plum
Steed-Kisker . . .	Platte	Mo.	115 1547-1557	. . . .	corn, squash, sunflower, walnut, hazelnut, pecan
Campbell Island	Butler	Ohio	121 1569	. . . .	sunflower

TABLE *(CONT.)*

Site	County	State	UMMAEL Rept. No. and Lab.No.	UMMA Cat. Nos.	Plant Remains
"Aylmer"*	Elgin	Ont.	176 2663-2714	. . . .	corn, beans, squash, walnut, acorn, butternut, beechnut
Castle Creek. . .	Broome	N.Y.	186 2842-2843	. . . .	corn, acorn
Baum.	Ross	Ohio	209 3172	. . . .	corn
Gartner	Ross	Ohio	209 3173	. . . .	corn
Feurt.	Scioto	Ohio	209 3174	. . . .	corn
Baldwin	Fairfield	Ohio	209 3174	. . . .	corn
Campbell Island.	Butler	Ohio	209 3176	. . . .	corn, beans
Norton*	Macomb	Mich.	228 3333-3335	. . . .	butternut
Reckner*	Somerset	Pa.	229 3356-3360	. . . . .	corn, chenopod, walnut, hickory nut
Martin	Fayette	Pa.	229 3361	. . . .	corn
Emerick	Somerset	Pa.	229 3362	. . . .	corn, beans
Powell 1	Somerset	Pa.	229 3363	. . . .	corn
Phillips*	Fayette	Pa.	229 3364-3368	. . . .	corn, hickory nut
Utz	Saline	Mo.	239 3400-3441	. . . .	corn, beans, squash, walnut, paw-paw, hazelnut, hackberry, cherry, plum
Belcher	Caddo	La.	240 3432-3435	. . . .	corn, beans, hickory nut, walnut
Lyons Farm . . .	Bradford	Pa.	246 3465	. . . .	corn
Davis	Cherokee	Texas	252 3486-3499	. . . .	corn, hickory nut, acorn
Muntz*	Warren	Pa.	260 3537-3563	. . . .	walnut, butternut, hickory nut
Cowan Creek* . .	Clinton	Ohio	263 3800-3801	. . . .	gourd, chenopod (Adena)
Seed	York	Ont.	288 3804	. . . .	corn
Krieger	Kent	Ont.	289 3805-3810	. . . .	corn, grape
Jones Village . .	Blue Earth	Minn.	292A 3919-3920	. . . .	corn
Cannon Junction .	Goodhue	Minn.	292B 3921	. . . .	corn

TABLE *(CONT.)*

Site	County	State	UMMAEL Rept. No. and Lab.No.	UMMA Cat. Nos.	Plant Remains
Angel	Vanderburgh	Ind.	293 3922-3930	....	corn, beans
Cahokia	St. Clair	Ill.	302 3996-4011	....	corn, squash
Goessens	Oxford	Ont.	303 4012	....	corn, chenopod
Ansell	Calhoun	Ill.	304 4013	....	corn (Hopewell?)
Dutton's Ash Pit	....	Ont.	313 4187-4190	....	corn, beans, Canada plum
Gibraltar	Wayne	Mich.	324 4363	24554	corn
Goldsmith	....	Ohio	333 4435	....	blackberry or raspberry
Canter's Caves .	Jackson	Ohio	368 4644-4645	....	corn
Kettle Hill Cave.	Fairfield	Ohio	368 4646-4647	....	corn
Fanning	Doniphan	Kans.	372 4652-4653	....	corn, groundnut
Rock Creek....	Adams	Pa.	390 4710	....	corn
20 MB 33	Macomb	Mich.	403 4755-4757	....	beans, hickory nut
Banks	Crittenden	Ark.	405 4759	....	corn
Serpent Mound* .	....	Ont.	409 4791	....	Canada plum
Kingston*	Peoria	Ill.	411 4794	....	corn, butternut
Varner	Greene	Pa.	412 4795	....	corn
Mikado Earthwork	Alcona	Mich.	413 4796	....	corn
Warren Dune...	Berrien	Mich.	414 4797	....	corn
20 IR 1/8*	Isle Royale	Mich.	415 4798	....	bearberry
Stroebel*	Saginaw	Mich.	415 4799	....	acorn
Hodges*	Saginaw	Mich.	415 4800	....	chenopod (modern?)
20 MR 5*	Monroe	Mich.	415 4801	....	corn, hickory nut
Verchave II* ...	Macomb	Mich.	415 4802	....	corn, hickory nut, acorn, grape, thornapple, walnut

TABLE *(CONT.)*

Site	County	State	UMMAEL Rept. No. and Lab.No.	UMMA Cat. Nos.	Plant Remains
Spoonville*	Ottawa	Mich.	415 4803	. . . .	butternut, walnut, acorn (Hopewell)
Schultz*	Saginaw	Mich.	415 4804	. . . .	acorn, walnut (Middle Woodland)
Green Point . . .	Saginaw	Mich.	. . . .	3324	corn
Indian Pt. I.R. .	Isle Royale	Mich.	. . . .	5367	hazelnut
"Mayville"	Chautauqua	N.Y.	. . . .	2403	corn
Moccasin Bluff*.	Berrien	Mich.	. . . .	29158	butternut
Ford Turf Garden	Wayne	Mich.	. . . .	39731	corn
"Ellington"	Chautauqua	N.Y.	. . . .	42188	corn

Holcombe Site (20 MB 30), Macomb County, Michigan*		
Catalogue Number	Square: Level	Wood Charcoal
58013.	570-470: Feat. 6	maple, tamarack, blue beech, sycamore
57924.	590-540: plow zone	birch, beech, chestnut, burned pitch
57925.	: 1-2	oak, chestnut
57928.	: 2-3	red oak
57936.	600-540: 1-2	red oak, pitch, beech
57937.	: 2	sycamore
57938.	: 1-2	burned pitch
57940.	: 2-3	sycamore
58135.	610-490: plow zone	pine, white oak, beech, poplar, sycamore
57996.	: 1-2	pine, beech
57942.	: 1-2	sycamore
57998.	: 1-2	pine, white oak
57993.	: 1-2	pine, beech
57943.	: 2-3	pine, beech, hickory
57945.	: 2-3	beech, black gum
57946.	: 2-3	pine, beech
57947.	610-500: plow zone	white oak, sycamore
57951.	: 1-2	pine, beech
57953.	: 2-3	pine, tulip tree
57958.	620-460: loose surface	beech, white oak group

TABLE *(CONT.)*

Catalogue Number	Square: Level	Wood Charcoal
57959	: plow zone	tulip tree
57955	620-460: 1	maple, beech
57960	: 2	beech, maple, white oak
57961	: 3	beech
57963	620-470: 1-2	sycamore, white oak, beech
57965	: 2-3	sycamore, red oak, willow, maple
57970	: 3-4	maple, red oak
57971	: 4	beech, maple
57972	620-480: B.S. 1-2	beech, maple
57973	: 1-2	beech, oak, pine
57974	: B.S. 1-2	beech
57979	: 2-3	beech, maple, oak
57981	: 2-3	red oak, maple
58000	: 2-3	pine, red oak
58015	: 3-4	*(carbonized chestnut fragment)*
58142	620-490: surface	beech, maple, sycamore
57990	: B.S. 1-2	beech, maple, pine
57992	: B.S. 1-2	red oak, pitch
58017	: B.S. 1-2	beech, burned pitch
58003	: B.S. 2-3	beech
58004	: B.S. 2-3	beech
58005	: B.S. 3-4	beech, maple
58012	: B.S. 3-4	beech
57984	:Feat. 3	(ragweed seed), beech, red and white oak
58045	620-500: surface	chestnut, sycamore, rock elm, maple, beech
58048	: plow zone	beech, tulip tree, pine, birch, sycamore, maple
58047	620-500: 1	pine
58051	:Feat. 2	pitch
58056	620-500: 1-2	maple, white pine
58052	: 2-3	beech, maple
58053	:Feat. 3	beech
58054	: 3	beech
58058	620-510: plow zone	rock elm, beech
58059	: 1-2	pitch, beech
58060	: 1-2	pitch, beech
58062	: 1-2	pitch, beech
58063	: 1-2	beech

TABLE *(CONT.)*

Catalogue Number	Square: Level	Wood Charcoal
58065.	: 2-3	beech, 1 piece basswood
58147.	630-470: 1-2	maple, beech, red oak
58148.	: 2-3	white oak, maple
57969.	: 3-4	maple, beech, red and white oak groups
58069.	630-480: plow zone	beech, maple
58072.	: 1-2	beech
58073.	: 1-2	beech
58074.	: 1-2	beech, sycamore, pine
58070.	: 2-3	white and red oak groups, beech
58076.	: 2-3	beech, rock elm
58077.	: 3, Feat. 7	beech
58078.	: 3	beech
58075.	: 3-4	beech, red oak group
58079.	630-480: 4	maple
58080.	: 5	ash
58081.	630-490: plow zone	ash
58085.	: 1	beech, birch
58086.	630-490: 1-2	beech
58087.	: 1-2	pine
58091.	: B.S. 1-2	tulip tree
58092.	: B.S. 1-2	maple, birch, tulip tree
58093.	: B.S. 2	beech, pine
58095.	: B.S. 2	pine
58096.	: B.S. 2	beech, red oak group
58110.	630-500: 1	maple, pine, tulip tree
58114.	: 1	beech
58115.	: 1-2	tulip tree
58116.	: 1-2	birch, beech
58119.	: 1-2	pine
58121.	: 2-3	beech
58122.	: 2-3	beech, pine
58123.	: 2-3	conifer
58124.	: 3-4	pine, beech
58125.	: 3-4	pine
58126.	: 3-4	pine
58127.	: 3-4	pine
58128.	630-510: plow zone	maple, beech, pine, white and red oak
58129.	: 1	pitch, pine, beech, sycamore

TABLE *(CONT.)*

Catalogue Number	Square: Level	Wood Charcoal
58130.	: 1-2	beech
58132.	630-510: 1-2	beech, pitch, maple
58133.	: 2-3	beech, pine
58134.	: 2-3	beech, burned pitch
58066.	640-490: plow zone	beech, 1 unidentified
58067.	895-420: plow zone	white oak, beech, 2 unidentified
58068.	: 1-2	white oak, beech

Feeheley Site (20 SA 128), Saginaw County, Michigan*

Field No. (Catalogue No.)	Square: Level	Feature	Seed and Nut Remains
212	-10-305:	. . . .	butternut
213	5-324: 0-1	. . . .	hickory nut
214	: 1	. . . .	butternut
214	: 1-2	. . . .	butternut, hickory
215	: 2-2.8'	. . . .	butternut
155	10-305: 1 and 2	. . . .	hickory nut
117	30-115: 1	. . . .	butternut
117	: below 1	. . . .	butternut
115	: below 2	. . . .	butternut
187	50-275: 3	. . . .	butternut
42	60-110: 0-1	. . . .	butternut
43	: 1-1.5'	. . . .	acorn
43	: 1-2	. . . .	butternut, walnut
43	: 2.0'-2	. . . .	butternut, walnut
96	60-130: 3	J	acorn
56	70-110: 2-3	B	acorn
32	70-120: 7	AE	butternut
136	70-400: 1-2	. . . .	butternut
65	80-110: 4-5	E	hickory nut, acorn
(56964)	90-110:	6	acorn
(50969)		0	acorn
17	90-120: 1	F	acorn
232	100-140: 1	. . . .	butternut
235	: 1	. . . .	grape seed

TABLE *(CONT.)*

Field No. (Catalogue No.)	Square: Level	Feature	Seed and Nut Remains (Wood Charcoal)
209	5-315:	. . . .	white oak, black oak, ash, beech
214	5-324: 1	. . . .	white oak, black oak, beech (1.4')
150	10-305: 1A	. . . .	pine (1.4'-2.2')
151	: 1-2	. . . .	black ash, beech, white oak, sycamore, black oak, pine (3')
170	25-305: 1	. . . .	pine
117	30-110:	. . . .	pine, ash, sycamore (1.8')
117	30-115: 1	. . . .	black oak, blue beech
161	30-300: 1-2'	. . . .	beech, sycamore, hickory, ash
163	: 2	Y	pine, beech (blackened area B)
164	: 2	Y	beech, ash, white elm, black oak (pit A)
165	: 2	Y	beech, ash, blue beech, black oak, white oak
166	: 2	Y	black oak
167	: 2	. . . .	black oak
118	40-115: 0.7'-1	. . . .	black oak, white oak, pine
218	40-350: 2	. . . .	black oak, pine (0.7')
219	: 3	. . . .	white oak (1')
199	55-250: 2	AN	black oak
40	60-120: 1-2	J	white oak, black oak, sycamore
38	60-120: 2	J?	black oak
41	: 2-3	J	black oak, white oak, beech, basswood
95	60-130: 2	N	sycamore, beech, sugar maple (fire pit)
93	: 2-3	. . . .	sycamore, beech, maple
96	: 3	J	ash, beech, sycamore, white oak, oak bark
97		4	sycamore, oak (burial)
105	60-150: 1	R	pine
87	65-125: 3-4	. . . .	white oak, beech, sycamore
203	65-190: 0-0.5'	. . . .	pine
31	70-120: 2-3	C	ash
32	: 7	AE	black ash, white ash, black oak, white oak, pine

TABLE *(CONT.)*

Field No. (Catalogue No.)	Square: Level	Feature	Wood Charcoal
134	70-140: 2	L	black ash, white ash, black oak, white oak, hickory
102		3	white oak, black oak, black ash, pine (burial)
44	70-400: 1-2	M?	pine, white oak, black oak, ash
144	: 6 +	M	black ash, white oak, sycamore
60	80-110: 1-2	. . . .	black oak, black gum (?)
65	: 3-4	D	oak bark, white oak, black oak
62	: 4-5	. . . .	ash
65	80-110: 4-5	E	white oak, black oak, white ash, white elm, slippery elm
76	80-120: 3-4	E	white oak, black oak
7	90-110:	8	black oak
19	90-120: 2-3	P	black oak, beech
232	100-140: 1	. . . .	white oak, black ash, pine, beech
132	100-445:	V	white oak, black oak, hickory
235	110-140: 1	. . . .	pine, beech

Juntunen Site (20 MK 1), Mackinac County, Michigan*

Square: Level	Plant Remains (seeds, nuts, tubers, and birchbark)
1600-1070: 11-12	birchbark
1600-1130: 3-4	corn, cherry (41535)
: 5-6	tuber, birchbark (41537)
: 5-6	acorn, blackberry, *Galium* bearberry, fire cherry, elder berry
: 8-9	birchbark
: 10-11	birchbark
1605-1130: 11-12	hazelnut
1680-1240: 3-4	hazelnut
1680-1245: 2-9	tuber
1700-1120: 1-2	hazelnut
: 0-5	corn, birchbark
: 4-5	corn, acorn, fire cherry, birchbark

TABLE *(CONT.)*

Square: Level	Plant Remains (seeds, nuts, tubers, and birchbark)
: 5-6	birchbark
: 5-6	corn
: 5-6	corn, hazelnut, Canada plum
: 5-6	Canada plum, birchbark
: 6-7	corn, hazelnut, birchbark
: 7-8	corn
: 7-8	birchbark
: 7-8	corn, hazelnut, Canada plum
: 8-9	corn, hazelnut, blueberry, tuber, birchbark
: 8-9	tuber
: 9-10	corn
: 10-11	birchbark
: 11-12	birchbark
: 12	birchbark
1700-1140:	corn
: 2-3	"large seed"
: 3-4	hazelnut, basswood bark
: 4-5	chenopod, elderberry, blackberry, fire cherry, *Galium* (Feature 20)
: 4-5	blackberry (Feature 20, area F)
: 4-5	fire cherry, bearberry, elderberry, *Galium* (Feature 20, area L)
: 4-5	fire cherry, bearberry, hazelnut, *Galium* (Feature 20, area I)
1700-1160: 1-2	fire cherry, chenopod *(Chenopodium hybridum* var. *gigantospermum)*
: 1-2	blackberry, chenopod *(Chenopodium hybridum* var. *gigantospermum)*
: > 5	hazelnut, sumac berry
1700-1180: 5-6	fire cherry, elderberry ', *Galium* (Bone Pit I)
1700-1190: 2-3	corn, "large seed"
: 3-4	birchbark
: 5-6	birchbark
: 6-7	birchbark
: 7-8	birchbark, "large seed"
1700-1200: 1-2	birchbark
: 4	hazelnut
1700-1200: 7-8	acorn, blackberry, elderberry, grape, fire cherry, chenopod, *Galium*
: 7-10	acorn, hazelnut
: 13-14	hazelnut

TABLE *(CONT.)*

Square: Level	Plant Remains (seeds, nuts, tubers, and birchbark)
: 13-14	fire cherry (Feature 13)
: 14-15	pepperroot
: 15-16	corn
1700-1210: 3-4	tuber
1710-1140: 4-5	hazelnut
1710-1180: 3-4	corn, birchbark
: 4	corn (Feature 21)
: 7-8	hazelnut, birchbark
1710-1180: 8-9	fire cherry
: 9-10	birchbark
1710-1210: 0-1	"large seed"
: 1-2	"large seed", hazelnut (not carbonized, probably intrusive)
: 4-5	hazelnut
: > 4	birchbark
1720-1210:	hazelnut (postmold 10 L)
: 4-5	pepperroot, fire cherry, bearberry, blackberry, elderberry, sumac berry, beechnut (hearth; 41618, 41619)
: 5-6	pepperroot, fire cherry, blackberry, birchbark
: > 5	fire cherry, elderberry
: 5-13	pepperroot, birchbark
: > 12	pepperroot
: 13-14	pepperroot
: 14	pepperroot, cherry (pit)
1740-1260: 2	tuber (0.4'-0.7')
1790-1135: 1-2	tuber (0.5'-1.0')
: 2-4	corn
1790-1220: 2-3	tuber (0.5'-1.0')
	tuber, hazelnut, birchbark (1.0'-1.5')
1870-1100: 8-9	corn
1870-1140: 1-2	hazelnut, birchbark
: 2-3	corn, fire cherry, blackberry, bearberry, hazelnut, birchbark, *Galium*
: 11-12	hazelnut
1880-1130: 4-5	hazelnut, birchbark
: 6-7	hazelnut, birchbark (184)
: 9-10	hazelnut
1890-1190: 5	corn, acorn, fire cherry (5 seeds), chenopod (430 seeds), blackberry (15 seeds), elderberry (30 seeds), sumac berry, beechnut, *Galium*, birchbark (Feat. 9; 41120)

TABLE *(CONT.)*

Square: Level	Plant Remains (seeds, nuts, tubers, and birchbark)
1890-1200: 5	fire cherry, blackberry, elderberry, chenopod, *Galium* (Feature 3)
: 6	corn, fire cherry, elderberry, blackberry, sumac berry (postmold no. 5)

LITERATURE CITED

Aller, Wilma F.
1954 Aboriginal Food Utilization of Vegetation by the Indians of the Great Lakes Region as Recorded in the Jesuit Relations. Wisconsin Archeologist, n.s., 35 (3): 59-73.

Anderson, Edgar
1940 Introgressive Hybridization. Biological Research Series. New York: John Wiley & Sons, Inc.
1952 Plants, Man and Life. Boston: Little, Brown & Co.
1956 Man as a Maker of New Plants and New Plant Communities. *In:* Man's Role in Changing the Face of the Earth, ed. William L. Thomas. Chicago: The University of Chicago Press.

Baerreis, David A. and Robert Nero
1956 The Storage Pits of the Dietz Site (Da 12) and Their Contents. Wisconsin Archeologist, 37 (1): 5-18.

Bailey, L. H.
1935 The Standard Cyclopedia of Horticulture. Vol. II. New York: The Macmillan Co.

Bakeless, John
1950 The Eyes of Discovery. Philadelphia: J.B. Lippincott Company.

Bank, Ted P. II
1953 Ecology of Prehistoric Aleutian Village Sites. Ecology, 34 (2): 246-64.

Barrett, S. A.
1933 Ancient Aztalan. Bulletin of the Public Museum of the City of Milwaukee, 8: 1-602.

Beardsley, Gretchen
1940 The Groundnut as Used by the Indians of Eastern North America. Papers of the Michigan Academy of Science, Arts, and Letters, 25: 507-15.

Bennington, Frederick, Carl Merton, and Patty Jo Watson
1962 Carbon Dating Prehistoric Soot from Salts Cave, Kentucky. American Antiquity, 28 (2): 238-41.

Biel, E. R.
1961 Microclimate, Bioclimatology and Notes on Comparative Dynamic Climatology. American Scientist, 49 (3): 326-57.

Billington, Cecil
1943 Shrubs of Michigan. Cranbrook Institute of Science, Bulletin 20.

Black, M. J.
1963 The Distribution and Archaeological Significance of the Marsh Elder, *Iva annua* L. Papers of the Michigan Academy of Science, Arts, and Letters, 48: 541-47.

Blair, E. H.
1911 Indian Tribes of the Upper Mississippi and the Great Lakes Region. 2 vols. Cleveland: The Arthur H. Clark Company.

Blankinship, J. W.
1905 Native Economic Plants of Montana. Montana Agricultural Experiment Station, Bulletin 56.

Boyle, David
1900 Archaeological Report. *In:* Appendix to the Report of the Minister of Education, Ontario for 1899. Toronto.

Bromley, S. W.
1945 An Indian Relict Area. Scientific Monthly, 60 (Feb.): 153-54.

Brown, Clair A.
n.d. Ferns and Flowering Plants of Isle Royale, Michigan. U. S. Department of the Interior.

Brown, W. L. and Edgar Anderson
1947 The Northern Flint Corns. Annals of the Missouri Botanical Garden, 34 (1): 1-28.
1948 The Southern Dent Corns. *Ibid.*, 35 (3): 255-68.

Byers, Douglas S.
1946 The Environment of the Northeast. *In:* Man In Northeastern North America, ed. Frederick Johnson. Papers of the Robert S. Peabody Foundation for Archaeology, 3: 3-32.

Cain, Stanley A. and J. V. Slater
1948 Palynological Studies at Sodon Lake, Michigan, Part III. The Sequence of Pollen Spectra, Profile I. Ecology, 29 (4): 492-500.

Canada, Department of Mines
1957 Atlas of Canada. Canada, Department of Mines, Technological Surveys, Geographical Branch. Ottawa.

Carr, Lucien
1895 The Food of Certain American Indians and Their Methods of Preparing It. American Antiquarian Society, Proceedings, n.s., 10 (1): 155-90.

Carter, George F. and Edgar Anderson
1945 A Preliminary Survey of Maize in the Southwestern United States. Annals of the Missouri Botanical Garden, 32 (3): 297-322.

Chamberlain, L. S.
1901 Plants Used by the Indians of Eastern North America. American Naturalist, 35 (409): 1-10.

Clements, F. E. and R. W. Chaney
1936 Environment and Life in the Great Plains. Carnegie Institution of Washington Supplementary Publications, No. 24.

Crane, H. R.
1956 University of Michigan Radiocarbon Dates I. Science, 124 (3224): 664-72.

Crane, H. R. and J. B. Griffin
1958*a* University of Michigan Radiocarbon Dates II. *Ibid.*, 127 (3306): 1098-105.
1958*b* University of Michigan Radiocarbon Dates III. *Ibid.*, 128 (3332): 1117-23.
1959 University of Michigan Radiocarbon Dates IV. American Journal of Science, Radiocarbon Supplement, 1: 173-98.
1960 University of Michigan Radiocarbon Dates V. *Ibid.*, 2: 31-48.
1961 University of Michigan Radiocarbon Dates VI. *Ibid.*, 3: 105-25.
1962 University of Michigan Radiocarbon Dates VII. *Ibid.*, 4: 183-203.
1963 University of Michigan Radiocarbon Dates VIII. *Ibid.*, 5: 228-53.

Curtis, J. T. and M. L. Partch
1948 Effect of Fire on the Competition Between Blue Grass and Certain Prairie Plants. American Midland Naturalist, 39 (2): 437-43.

Cutler, Hugh C.
1956 Corn From the Dietz Site, Dane County, Wisconsin. Wisconsin Archeologist, 37 (1): 18-19.

Cutler, Hugh C. and Thomas W. Whitaker.
1961 History and Distribution of the Cultivated Cucurbits in the Americas. American Antiquity, 26 (4): 469-85.

Day, Gordon M.
1953 The Indian as an Ecological Factor in the Northeastern Forest. Ecology, 34 (2): 329-46.

Deale, Valentine B.
1958 The History of the Potawatomis Before 1722. Ethnohistory, 5 (4): 305-60.

Deam, Charles C.
1940 Flora of Indiana. Indianapolis: Indiana Department of Conservation.

Densmore, Frances
1928 Uses of Plants by the Chippewa Indians. Bureau of American Ethnology, Annual Report, 44: 275-397.
1929 Chippewa Customs. Bureau of American Ethnology, Bulletin 86.

Dice, L. R.
1938 The Canadian Biotic Province with Special Reference to the Mammals. Ecology, 19 (4): 503-14.

Douglas, F. H.
1941 An Ottawa Cedar Bark Mat. Denver Art Museum, Department of Indian Art, Material Culture Notes, No. 16.

Driver, Harold E. and William C. Massey
1957 Comparative Studies of North American Indians. Transactions of the American Philosophical Society, n.s., 47 (2): 163-456.

Driver, Harold E., John M. Cooper, Paul Kirchoff, Dorothy Ranier Libby, William C. Massey, and Leslie Spier
1953 Indian Tribes of North America. International Journal of American Linguistics, Memoir 9.

Espenshade, Edward B., Jr. (ed.)
1962 Rand McNally Regional Atlas, 2d ed.; Chicago: Rand McNally & Co.

Fenton, William N.
1940 An Herbarium from the Allegany Seneca. *In:* The Historic Annals of Southwestern New York, pp. 787-96. New York: Lewis Historical Publishing Co., Inc.

Fernald, M. L.
1950 Gray's Manual of Botany, 8th ed.; New York: American Book Company.

Fernald, M. L. and A. C. Kinsey
1943 Edible Wild Plants of Eastern North America. Gray Herbarium of Harvard University, Special Publication. Cornwall-on-Hudson: Idlewild Press.

Finch, V. C. and O. E. Baker
1917 Geography of the World's Agriculture. U. S. Department of Agriculture, Office of Farm Management. Washington: U. S. Government Printing Office.

Fink, Bruce and Joyce Hedrick
1935 The Lichen Flora of the United States. Ann Arbor: University of Michigan Press.

Flint, Richard F. and Edward S. Deevey (eds.)
1962 Radiocarbon. Vol. 4. New Haven: American Journal of Science.

Forest Service
1948 Woody-plant Seed Manual. U. S. Department of Agriculture Miscellaneous Publication 654. Washington: U. S. Government Printing Office.

Forestry Branch
1956 Native Trees of Canada. Canada Department of Northern Affairs and National Resources Bulletin 61. Ottawa: Queen's Printer.

Fowler, Melvin L.
1952 The Clear Lake Site: Hopewellian Occupation. Hopewellian Communities in Illinois. Illinois State Museum, Science Papers, 5 (4): 131-74.

Funkhouser, W. D. and W. S. Webb
1930 Rockshelters of Wolfe and Powell Counties, Kentucky. The University of Kentucky Reports in Archaeology and Anthropology, 1 (4): 233-306.

Galinat, Walton C. and James H. Gunnerson
1963 Spread of Eight-Rowed Maize from the Prehistoric Southwest. Harvard University, Botanical Museum Leaflets 20 (5): 117-60.

Geographical Branch
1957 Atlas of Canada. Department of Mines and Technical Surveys, Ottawa: Queen's Printer.

Gilmore, Melvin R.
1933 Some Chippewa Uses of Plants. Papers of the Michigan Academy of Science, Arts, and Letters, 17: 119-43.

Goodspeed, T. H.
1954 The Genus *Nicotiana:* Origins, Relationships and Evolution of its Species in the Light of Their Distribution, Morphology, and Cytogenetics. Chronica Botanica, 16 (1/6): 1-536. Waltham, Mass.: Chronica Botanica Company.

Goslin, Robert M.
1952 Cultivated and Wild Plant Food from Aboriginal Sites in Ohio. Ohio Archaeologist, 2 (2): 9-29.
1957 Food of the Adena People. *In:* The Adena People No. 2, Wm. S. Webb and Raymond S. Baby. Ohio Historical Society: The Ohio State University Press (41-46).

Greenman, Emerson F.
1937*a* Two Prehistoric Villages Near Cleveland, Ohio. Publications of the Ohio Archaeological and Historical Society, 46: 305-66.
1937*b* The Younge Site, an Archaeological Record from Michigan. Occasional Contributions from the Museum of Anthropology of the University of Michigan, No. 6. Ann Arbor.
1939 The Wolf and Furton Sites, Macomb County, Michigan. *Ibid.*, No. 8.
1961 The Indians of Michigan. Michigan Historical Commission, John M. Munson Michigan History Fund, Pamphlet 5.

Griffin, James B.
1943 The Fort Ancient Aspect. Ann Arbor: University of Michigan Press.
1946 Cultural Change and Continuity in Eastern United States Archaeology. *In:* Man in Northeastern North America, ed. Frederick Johnson. Papers of the Robert S. Peabody Foundation for Archaeology, 3: 39-95.
1949 Meso-America and the Southeast: A Commentary. *In:* The Florida Indian and his Neighbors, ed. John W. Griffin. Inter-American Center, Rollins College, Winter Park, Florida. pp. 77-100.
1952 The Late Prehistoric Cultures of the Ohio Valley. Ohio State Archaeological and Historical Quarterly, 61 (2): 186-95.
1958 The Chronological Position of the Hopewellian Culture in the Eastern United States. Anthropological Papers, Museum of Anthropology, University of Michigan, No. 12. Ann Arbor.
1960*a* Climatic Change: a Contributory Cause of the Growth and Decline of Northern Hopewell Culture. The Wisconsin Archaeologist, 14 (2): 21-33.
1960*b* A Hypothesis for the Prehistory of the Winnebago. *In:* Culture in History, Essays in Honor of Paul Radin, ed. Stanley Diamond. New York: Columbia University Press.
1961*a* Post-glacial Ecology and Culture Changes in the Great Lakes Area of North America. Great Lakes Research Division, Institute of Science and Technology, University of Michigan. Publication 7.

1961*b* Comment on Neolithic Diffusion Rates by Munro S. Edmonson. Current Anthropology, 2 (2): 92-93.

Griffin, James B. and Richard G. Morgan (editors)
1941 Contributions to the Archaeology of the Illinois River Valley. Transactions of the American Philosophical Society, 32 (1): 1-209.

Griffin, James B. and Richard A. Yarnell
1963 A New Radiocarbon Date on Corn from the Davis Site, Cherokee County, Texas. American Antiquity, 28 (3): 396-97.

Guthe, Alfred K.
1958 The Late Prehistoric Occupation in Southwestern New York: An Interpretive Analysis. New York State Archaeological Association, Researches and Transactions, 14: 1-100.

Hall, Robert L.
1962 The Archaeology of Carcajou Point, Vol. I. Madison: University of Wisconsin Press.

Hanes, Clarence R. and Florence N. Hanes
1947 Flora of Kalamazoo County, Michigan. Portland, Maine: The Anthoensen Press.

Harlow, William M.
1957 Trees of the Eastern and Central United States and Canada. New York: Dover Publications, Inc.

Harrington, M. R.
1908 Some Seneca Corn Foods and Their Preparation. American Anthropologist, 10 (4): 575-90.

Harris, George H.
1891 Root Foods of the Seneca Indians. Rochester Academy of Science, Proceedings, 1 (Dec.): 106-17.

Havard, V.
1895 Food Plants of the North American Indians. Bulletin of the Torrey Botanical Club, 22 (3): 98-123.

Hedrick, U. P.
1933 A History of Agriculture in the State of New York. Albany: J. B. Lyon Co.

Heizer, Robert F.
1955 Primitive Man as an Ecologic Factor. The Kroeber Anthropological Society, Papers, 13: 1-31.

Henshaw, H. W.
1890 Indian Origin of Maple Sugar. American Anthropologist, o.s., 3 (2): 341-51.

Hickerson, Harold
1962 The Southwestern Chippewa, an Ethnohistorical Study. American Anthropological Association, Memoir 92.

Hinsdale, Wilbert B.
1928 Indian Corn Culture in Michigan. Papers of the Michigan Academy of Science, Arts, and Letters, 8: 31-49.

1931 Archaeological Atlas of Michigan. Michigan Handbook Series, No. 4. University Museums, University of Michigan. Ann Arbor: University of Michigan Press.

1932 Distribution of the Aboriginal Population of Michigan. Museum of Anthropology, University of Michigan, Occasional Contributions, No. 2.

Hoffman, W. J.
1896 The Menomini Indians. Bureau of American Ethnology, Annual Report 14: 3-328.

Hunt, George T.
1940 The Wars of the Iroquois. Madison: University of Wisconsin Press.

Hunter, John D.
1823*a* Memoirs of a Captivity Among the Indians of North America. London: Longman & Co.
1823*b* Manners and Customs of Several Indian Tribes West of the Mississippi. Philadelphia: printed and published for the author by J. Maxwell.

Jenks, A. E.
1900 Wild Rice Gatherers of the Upper Lakes. Bureau of American Ethnology, Annual Report, 19: 1011 1161.

Johnston, Alexander
1962 *Chenopodium album* as a Food Plant in Blackfoot Indian Prehistory. Ecology, 43 (1): 129-30.

Jones, Peter R.
1861 History of the Ojibway Indians. London: A. W. Bennett.

Jones, Volney H.
1936*a* Some Chippewa and Ottawa Uses of Sweet Grass. Papers of the Michigan Academy of Science, Arts, and Letters, 21: 21-31.
1936*b* The Vegetal Remains of Newt Kash Hollow Shelter. *In:* Rockshelters in Menifee County, Kentucky, W. S. Webb and W. D. Funkhouser.
1937 Notes on the Preparation and Uses of Basswood Fiber by the Indians of the Great Lakes Region. Papers of the Michigan Academy of Science, Arts, and Letters, 22: 1-14.
1948 Notes on the Manufacture of Cedar-bark Mats by the Chippewa Indians. *Ibid.*, 32: 341-63.
1949 Maize from the Davis Site: Its Nature and Interpretation. *In:* The George C. Davis Site, Cherokee County, Texas, H. Perry Newell and Alex D. Krieger. Society for American Archaeology, Memoir 5. Pp. 241-49.

Jones, Volney H. and Robert L. Fonner
1954 Plant Materials from Sites in the Durango and La Plata Areas, Colorado. Basket Maker II Sites Near Durango, Colorado. Earl H. Morris and Robert F. Burgh. Carnegie Institution of Washington, Publication 604. Pp. 93-115.

Jury, Wilfred
1941 The Clearville Prehistoric Village Site in Oxford Township, Kent County, Ontario. Bulletin of the Museum, University of Western Ontario, No. 2. London, Ont., Canada.

Kendrew, W. G.
1942 The Climates of the Continents. 3d ed.; New York: Oxford University Press.

Kidd, Kenneth E.
1952 Sixty Years of Ontario Archeology. Archeology of the Eastern United States, ed. James B. Griffin. Chicago: The University of Chicago Press. Pp. 71-82.

Kinietz, W. Vernon
1940 The Indians of the Western Great Lakes, 1615-1670. Occasional Contributions from the Museum of Anthropology of the University of Michigan, No. 10. Ann Arbor.

Kinietz, W. Vernon and Volney H. Jones
1942 Notes on the Manufacture of Rush Mats Among the Chippewa. Papers of the Michigan Academy of Science, Arts, and Letters, 27: 525-37.

Kinsey, W. Fred
1960 The Sheep Rock: A Dry Shelter in Central Pennsylvania. Eastern States Archaeological Federation, Bulletin, 19: 17.

Kivett, Marvin F.
1952 Woodland Sites in Nebraska. Nebraska State Historical Society, Publications in Anthropology, No. 1.

Kroeber, A. L.
1939 Cultural and Natural Areas of Native North America. University of California Papers in American Archaeology and Ethnology 38: 1-242.

Lee, Thomas E.
1958 The Parker Earthwork, Corunna, Ontario. Pennsylvania Archaeologist, 28: 5-32.

Libby, W. F.
1952 Radiocarbon Dating. Chicago: The University of Chicago Press.
1954*a* Chicago Radiocarbon Dates IV. Science, 119 (3083): 135-40.
1954*b* Chicago Radiocarbon Dates V. *Ibid.*, 120 (3123): 733-42.

Little, Elbert L., Jr.
1949 Important Forest Trees of the United States. *In:* Trees, the Yearbook of Agriculture. U. S. Department of Agriculture, Washington: U. S. Government Printing Office. Pp. 763-814.

McCallum, K. J. and W. Dyck
1960 University of Saskatchewan Radiocarbon Dates II. American Journal of Science, Radiocarbon Supplement, 2: 73-81.

McClendon, J. F.
1921 Some American Plants Considered as Sources of Vitamins, and as Parts of a Diet Favorable to the Preservation of Teeth. The Journal of Dental Research, 3 (2): 279-95.

McGregor, John C.
1958 The Pool and Irving Villages, A Study of Hopewell Occupation in the Illinois River Valley. Urbana: University of Illinois Press.

Macpherson, J. C.
1878 Observations on the Pre-historic Earthworks of Wayne County, Indiana. Geological Survey of Indiana, Annual Reports, 8-10: 219-26.

McKern, W. C.
1945 Preliminary Report on the Upper Mississippi Phase in Wisconsin. Bulletin of the Public Museum of the City of Milwaukee, 16 (3): 109-285.

Mangelsdorf, Paul C. and R. G. Reeves
1939 The Origin of Indian Corn and its Relatives. Texas Agricultural Experiment Station, Bulletin 574.

Marks, William D.
1961 Michigan and the Great Lakes. Lansing: Michigan Water Resources Commission.

Maxwell, Hu
1910 The Use and Abuse of Forests by the Virginia Indians. William and Mary Quarterly Historical Magazine, 19 (1): 73-103.

Mayer-Oakes, Wm. J.
1955 Prehistory of the Upper Ohio Valley; an Introductory Archaeological Study. Annals of the Carnegie Museum. Vol. 34. Anthropological Series, No. 2.

Meighan, C. W., D. M. Pendergast, B. K. Swartz, Jr., and M. D. Wissler
1958 Ecological Interpretation in Archaeology: Part II. American Antiquity, 24 (2): 131-50.

Medsger, Oliver P.
1939 Edible Wild Plants. New York: The Macmillan Co.

Merrill, E. D.
1954 The Botany of Cook's Voyages and Its Unexpected Significance in Relation to Anthropology, Biogeography and History. Chronica Botanica, 14 (5/6): 161-384. Waltham, Mass.: Chronica Botanica Company.

Michigan Writers' Program
1941 Michigan, a Guide to the Wolverine State. New York: Oxford University Press.

Morse, Dan and Phyllis
1960 A Preliminary Report on 9-G0-507: The Williams Site, Gordon County, Georgia. The Florida Anthropologist, 13 (4): 81-91, 109-14.

Morse, Jedediah
1822 A Report to the Secretary of War of the United States on Indian Affairs. New Haven: S. Converse.

Mosely, E. L.
1931 Some Plants That Were Probably Brought to Northern Ohio from the West by the Indians. Papers of the Michigan Academy of Science, Arts, and Letters, 13: 169-72.

Murdock, George P.
1960 Ethnographic Bibliography of North America. 3d ed.; New Haven: Human Relations Area Files.

Nearing, G. G.
1947 The Lichen Book. Ridgewood, N.J.: G. G. Nearing.

Nelson, N. C.
1917 Contributions to the Archaeology of Mammoth Cave and Vicinity, Kentucky. American Museum of Natural History, Anthropological Papers, 22 (1): 1-73.

Newberry, J. S.
1887 Food and Fiber Plants of the North American Indians. Popular Science Monthly, 32 (Nov.): 31-46.

Newell, H. Perry and Alex D. Krieger
1949 The George C. Davis Site, Cherokee County, Texas. Society for American Archaeology, Memoir 5.

Nichols, G. E.
1935 The Hemlock-White Pine-Northern Hardwood Region of Eastern North America. Ecology, 16 (3): 403-22.

Oehler, Charles M.
1950 Turpin Indians. Cincinnati Museum of Natural History. Popular Publications Series, No. 1.

Otis, Charles H.
1931 Michigan Trees. Ann Arbor: University of Michigan Press.

Palmer, Edward
1871 Food Products of the North American Indians. Report of the U. S. Commissioner of Agriculture for 1870, pp. 404-28.

Parker, Arthur C.
1907 Excavations in an Erie Indian Village and Burial Site at Ripley, Chautauqua Co., N. Y. New York State Museum Bulletin 117.
1908 Report on the Archaeology Section. Fourth Report of the Director of the Science Division. *Ibid.*, 121: 85-110.
1910 Iroquois Uses of Maize and Other Food Plants. *Ibid.*, No. 144.
1918 A Prehistoric Iroquoian Site on the Reed Farm, Richmond Mills, Ontario County, N.Y. Researches and Transactions of the New York State Archaeological Association 1: 1-41.

Peterson, Maude G.
1923 How to Know Wild Fruits. New York: The Macmillan Co.

Potzger, J. E.
1946 Phytosociology of the Primeval Forest in Central-Northern Wisconsin and Upper Michigan, and a Brief Post-Glacial History of the Lake Forest Formation. Ecological Monographs, 16 (3): 211-50.

Preston, Richard S., Elaine Person, E. S. Deevey
1955 Yale Natural Radiocarbon Measurements II. Science, 122 (3177): 954-60.

Putnam, F. W.
1886 Eighteenth Report of the Curator. Peabody Museum of American Archaeology and Ethnology. *In:* Eighteenth and Nineteenth Annual Reports of the Trustees. Pp. 401-32.

Quimby, George I., Jr.
1939 Aboriginal Camp Sites on Isle Royale, Michigan. American Antiquity, 4 (3): 215-23.
1941 The Goodall Focus, an Analysis of Ten Hopewellian Components in Michigan and Indiana. Indiana Historical Society, Prehistory Research Series, 2: 61-161.
1960 Indian Life in the Upper Great Lakes. Chicago: The University of Chicago Press.
1961 The Pic River Site. *In:* Lake Superior Copper and the Indians: Miscellaneous Studies of Great Lakes Prehistory, ed. James B. Griffin. Anthropological Papers, Museum of Anthropology, University of Michigan, No. 17. Ann Arbor. Pp. 83-89.

Radin, Paul
1923 The Winnebago Tribe. Bureau of American Ethnology, Annual Report, 37: 35-550.

Radisson, Peter E.
1885 Voyages—Being an Account of His Travels and Experiences Among the North American Indians from 1652 to 1684. Boston: Prince Society.

Reagan, Albert B.
1928 Plants Used by the Bois Fort Chippewa. Wisconsin Archaeologist, n.s., 7 (4): 230-48.

Ritchie, William A.
1944 The Pre-Iroquoian Occupations of New York State. Rochester Museum of Arts and Sciences, Memoir 1.
1945 An Early Site in Cayuga County, New York, Type Component of the Frontenac Focus, Archaic Pattern. Rochester Museum of Arts and Sciences, Research Records 7.
1959 Excavation of an Owasco Village Site in New York: Report on 1958 Settlement Pattern Studies in the Northeast. Eastern States Archeological Federation, Bulletin No. 18: 11-12.

Roedl, Leo J. and James H. Howard
1957 Archaeological Investigations at the Renner Site. The Missouri Archaeologist, 19: 53-90.

Rogers, Edward S.
1962 The Round Lake Ojibwa. Art and Archaeology Division, Royal Ontario Museum, University of Toronto, Occasional Paper 5.

Rubin, Meyer and Corrine Alexander
1960 U. S. Geological Survey Radiocarbon Dates V. American Journal of Science, Radiocarbon Supplement, 2: 129-85.

Rubin, Meyer and Sarah M. Berthold
1961 U. S. Geological Survey Radiocarbon Dates VI. American Journal of Science, Radiocarbon, 3: 86-98.

Sauer, Carl O.
1952 Agricultural Origins and Dispersals. New York: The American Geographical Society.

Schorger, A. W.
1937 The Range of the Bison in Wisconsin. Transactions of the Wisconsin Academy of Science, Arts, and Letters, 30: 117-30.

Sears, Paul B.
1942 Postglacial Migration of Five Forest Genera. American Journal of Botany, 29 (8): 684-91.

Shantz, H. L. and Raphael Zon
1924 Natural Vegetation. *In:* Atlas of American Agriculture, O. E. Baker. Washington: U. S. Government Printing Office.

Skinner, Alanson
1921 Material Culture of the Menomini. Museum of the American Indian, Indian Notes and Monographs, No. 20.
1925 Observations on the Ethnology of the Sauk Indians. III. Notes on Material Culture. Bulletin of the Public Museum of the City of Milwaukee, 5 (3): 119-80.
1926 The Mascoutens or Prairie Potawatomi. II. Notes on Material Culture. *Ibid.*, 6 (2): 263-326.

Small, John K.
1927 Among Floral Aborigines. Journal of the New York Botanical Garden, 28: 1-2, 25-40.
1933 Manual of the Southeastern Flora. New York: J. K. Small.

Smith, Huron H.
1923 Ethnobotany of the Menomini Indians. Bulletin of the Public Museum of the City of Milwaukee, 4 (1): 1-174.
1928 Ethnobotany of the Meskwaki Indians. *Ibid.*, 4 (2): 175-326.
1932 Ethnobotany of the Ojibwe Indians. *Ibid.*, 4 (3): 327-525.
1933 Ethnobotany of the Forest Potawatomi Indians. *Ibid.*, 7 (1): 1-230.

Spaulding, Albert C.
1955 Prehistoric Cultural Development in the Eastern United States. *In:* New Interpretations of Aboriginal American Culture History. Washington, The Anthropological Society of Washington. pp. 12-27.

Stickney, Gardner P.
1896 Indian Use of Wild Rice. American Anthropologist (o.s.), 9: 115-21.

Stout, A. B.
1914 Vegetal Foods of the American Indians. Journal of the New York Botanical Garden, 15: 50-60

Stowe, G. C.
1940 Plants Used by the Chippewa. Wisconsin Archeologist, n.s., 21 (1): 8-13.

Strong, William D.
1935 An Introduction to Nebraska Archaeology. Smithsonian Miscellaneous Collections, 93, No. 10.

Struever, Stuart
1962 Implications of Vegetal Remains from an Illinois Hopewell Site. American Antiquity, 27 (4): 584-87.

Tantaquidgeon, Gladys
1932 Notes on the Origin and Uses of Plants of the Lake St. John Montagnais. The Journal of American Folk-Lore, 45 (176): 265-67.

Thomas, Cyrus
1894 Report on the Mound Exploration of the Bureau of Ethnology. Bureau of American Ethnology, Annual Report, 12: 1-742.

Thompson, A. C.
1954 The Boyer's Run Rock Shelter, a Preliminary Report. Pennsylvania Archaeologist, 24 (3-4): 115-26.

Thwaites, Reuben G. (ed.)
1896-1901 The Jesuit Relations and Allied Documents. 73 vols. Cleveland: Burrows Brothers.

U. S. Department of Agriculture
1941 Climate and Man. Yearbook of Agriculture. Washington: U. S. Government Printing Office.
1961 Seeds. *Ibid.*

Vanderlaan, Stanley
1960 The Wilkins Site (Abn-1-1). The New York State Archaeological Association, Bulletin, 18: 5-6.

Veatch, J. O.
1959 Presettlement Forest in Michigan. Department of Resource Development, Michigan State University. East Lansing.

Vietzen, Raymond C.
1945 The Immortal Eries. Elyria, O.: Wilmot Printing Co.

Waugh, F. W.
1916 Iroquois Foods and Food Preparation. Canada Department of Mines, Geological Survey, Anthropological Series, No. 12.

Weatherwax, Paul
1954 Indian Corn in Old America. New York: The Macmillan Co.

Webb, W. S.
1952 The Archaic Cultures and the Adena People. Ohio State Archaeological and Historical Quarterly, 61 (2): 173-81.

Webb, William S. and Raymond S. Baby
1957 The Adena People, No. 2. Columbus: The Ohio State University Press.

Webb, W. S. and W. D. Funkhouser
1936 Rock Shelters in Menifee County, Kentucky. The University of Kentucky, Reports in Archaeology and Anthropology, 3. 101-67.

Wedel, Waldo R.
1938 Hopewellian Remains Near Kansas City, Missouri. U. S. National Museum, Proceedings, 86: 99-106.

1943 Archaeological Investigations in Platte and Clay Counties, Missouri. U. S. National Museum, Bulletin 183.
1959 An Introduction to Kansas Archaeology. Bureau of American Ethnology, Bulletin 174.

West, George A.
1934 Tobacco, Pipes and Smoking Customs of the American Indians. Bulletin of the Public Museum of the City of Milwaukee, 17: 1-994.

White, Marian E.
1961 Iroquois Culture History in the Niagara Frontier Area of New York State. Anthropological Papers, Museum of Anthropology, University of Michigan, No. 16. Ann Arbor.

Whitford, A. C.
1941 Textile Fibers Used in Eastern Aboriginal North America. American Museum of Natural History, Anthropological Papers, 38 (1): 1-22.

Will, George F.
1924 Archaeology of the Missouri Valley. *Ibid.*, 22 (6): 285-344.

Williams, Joseph E. (ed.)
1959 Prentice-Hall World Atlas. Englewood Cliffs, N.J.: Prentice-Hall, Inc.

Willoughby, Charles C.
1922 The Turner Group of Earthworks. Hamilton County, Ohio. Peabody Museum of American Archaeology and Ethnology, Harvard University, Papers, 8: 1-132.

Wintemberg, W. J.
1928 Uren Prehistoric Village Site, Oxford County, Ontario. National Museum of Canada, Bulletin 51: 1-97.
1936 Roebuck Prehistoric Village Site, Grenville County, Ontario. *Ibid.*, 83: 1-169.
1939 Lawson Prehistoric Village Site, Middlesex County, Ontario. *Ibid.*, 94: 1-104.
1946 The Sidey-Mackay Village Site. American Antiquity, 11 (3): 154-182.
1948 The Middleport Prehistoric Village Site. Canada Department of Mines, Geological Survey, Museum Bulletins, 109: 1-79.

Yarnell, R. A.
1958 Implications of Pueblo Ruins as Plant Habitats. Unpublished Master's thesis, University of New Mexico.

Young, Bennett R.
1910. The Prehistoric Men of Kentucky. Filson Club Publications, No. 25. Louisville.

Zook, L. L.
1915 Tests of Corn Varieties on the Great Plains. U. S. Department of Agriculture, Bulletin 307.

Zumberge, James H. and John E. Potzger
1956 Late Wisconsin Chronology of the Lake Michigan Basin Correlated with Pollen Studies. Geological Society of America, Bulletin, 67: 271-88.

www.ingramcontent.com/pod-product-compliance
Lightning Source LLC
LaVergne TN
LVHW050020250826
846302LV00008B/41